A

GEOLOGICAL MAP

OF THE

United States,

AND THE

BRITISH PROVINCES OF NORTH AMERICA;

WITH AN

Explanatory Text, Geological Sections,

AND

PLATES OF THE FOSSILS WHICH CHARACTERIZE THE FORMATIONS.

BY

JULES MARCOU,

UNITED STATES GEOLOGIST, MEMBER OF THE GEOLOGICAL SOCIETY OF FRANCE, ETC. ETC.

" La surface du globe terrestre, malgré son irregularité apparente, n'est pas dessinée au hasard comme les courbes de fantaisie d'un jardin anglais, mais elle a beaucoup plus d'analogie avec nos parcs à la française, tels que ceux de Versailles et de St. Cloud, dont l'ordonnance générale se rapporte à des lignes droites, connexes entre elles, et où les lignes sinueuses ne se montrent que dans les details." — *Notice sur les Systêmes de Montagnes*, L. ELIE DE BEAUMONT.

BOSTON:

GOULD AND LINCOLN,

59 WASHINGTON STREET.

1853.

STEREOTYPED BY
HOBART & ROBBINS,
NEW ENGLAND TYPE AND STEREOTYPE FOUNDERY,
BOSTON.

TO

Prof. Louis Agassiz,

AUTHOR OF "POISSONS FOSSILES," MEMBER OF THE ROYAL SOCIETY, LONDON, OF THE INSTITUTE OF FRANCE, ETC. ETC.

MY DEAR SIR:

IT is with great pleasure that I avail myself of your kind permission to dedicate to you this slight sketch of a Geological Map of a part of North America, and thus give to it the sanction of your extensive and profound knowledge in all the departments of Natural History.

Perhaps you yet remember that my first geological work, on the Jura, was published under your auspices, and that at your special request the administration of the Garden of Plants intrusted to me the mineral and geological exploration of this part of the New World. Accustomed, from my first steps in science, to your learned counsels and your extreme kindness, you will find in the following pages ideas and observations you have frequently communicated to me in our conversations on the Geology of this country. Besides, we have traversed together the rivers and the great lakes of the West, sheltered beneath the same tent in those wild regions, as yet unexplored, and inhabited by the untamed child of the virgin forest, the red warrior, the American man.

These recollections of our journeyings are here geologically reviewed ; and I shall be very happy if their perusal recalls to you the pleasant days we have passed together, in studying the geological formations of North America.

Believe me, dear sir, with great respect,

Your affectionately attached friend,

JULES MARCOU.

HARRISON SQUARE, DORCHESTER,
June, 1853.

CONTENTS.

CHAPTER I.

PALÆOZOIC ROCKS.

CHAPTER II.

SECONDARY ROCKS.

CHAPTER III.

CHAPTER IV.

APPENDIX.

INTRODUCTION.

The American Continent, the New World of Columbus, of Cabot, of Cartier, and other navigators, is to the eye of a geologist more ancient than Europe. It had already nearly its present form, while Europe was buried beneath the ocean, and mostly only represented by some scattered islands, in the space from England and Spain up to the confines of Russian Asia. The places where now are situated the cities of Washington, Quebec, New York, Boston, Cincinnati and St. Louis, had already emerged from the bosom of the ocean; while London, Paris, Brussels, Vienna and Berlin, remained buried for thousands of years longer under the waters of the sea, which then washed the base of the rocks that at this day support the Capitol, the monuments of Wolfe and Montcalm, and the astronomical observatory of Cincinnati.

At the period of the discovery of America, geology was not yet born; the spirit that presides over the progress of modern science was hardly awakened, and it was not until three centuries after the celebrated discovery of the Genoese navigator that geology began to escape from the clouds of fiction and mythology, and to take her proper rank as an exact science between physics and zoology.

The genius of Werner, of De Saussure, and of Cuvier, laid the foundations on which geology now rests. Those illustrious pioneers first opened the paths which we now follow with so much ease; and if in our days we are able, so to speak, to rebuild the continents, by recognizing the epoch at which each of the chains of mountains and the basins that constitute them were formed, let us not forget the long and difficult labors of our predecessors; the painful hesitations, and the study necessary to overcome them, before those masters of the science could reconstruct the fauna and flora of the earlier ages of our planet. One of the most distinguished generals and military writers of France has defined the art of war as being "the resumé of all human knowledge;" geology may with equal justice be styled the epitome of physical and natural science.

The first geological surveys in America were made from 1812 to 1816, by Lesueur and Maclure. The latter published in 1817 a geological map of all the original states of the Union which were east of the Alleghany Mountains. In 1825, the celebrated and unfortunate Capt. John Franklin, in company with John Richardson, published a geological journal of his second land journey for the discovery of the north-west passage, following the great lakes, and the rivers of the territory belonging to the Hudson's Bay Company. In the same year a young lieutenant of the English royal navy, now the learned hydrographer and geologist, Captain H. W. Bayfield, began his work on the geological constitution of the basin of the St. Lawrence, — a work which he has constantly pursued for twenty-seven years, and which is collected in a series of very interesting memoirs, published in the Proceedings of the Historical Society of Quebec, and the Geological Society of London.

Messrs. Conrad & Lea, celebrated conchologists of Philadelphia, published, in 1831 and 1833, an account of the numerous and interesting fossils which are found in the different sedimentary formations of America. Finally, Messrs. Jackson & Alger published, in 1828, a geological description of Nova Scotia and the Bay of Fundy, with a map and sections, of great interest.

Such, nearly, was the condition of geology in America when Mr. Murchison published in England his celebrated book entitled "*The Silurian System.*" This learned and elaborate work marks the commencement of a new era in those geological studies which relate to what had been called, until then, the Transition or Greywacke formation.

In 1835, a series of researches was commenced, which extended almost over the entire globe, in order to identify formations until then in a doubtful and uncertain stratigraphic position with those which Mr. Murchison had described on the borders of Wales. These

Silurian, Devonian, and Carboniferous formations, were recognized successively in Ireland, France, Belgium, on the borders of the Rhine, in Bohemia, Scandinavia, Russia, Spain, and Morocco. America was not slow in profiting by the beautiful discovery that Mr. Murchison had placed before the eyes of the learned world; and in 1835 Mr. Featherstonhaugh recognized the Silurian system on the borders of Lake Michigan, and in Wisconsin. Troost, Vanuxem and Eaton, were also among the first to compare the American formations with those of Europe; and laid the true foundations on which all the geological maps and memoirs published on this side of the Atlantic, for sixteen years, have been constructed.

Most of the state governments composing the American Union, and the federal government itself, recognizing the practical utility of geological maps and reports, made appropriations for the geological survey of the greater part of the country. The British Provinces followed this movement, and intrusted to men specially appointed the care of describing the rocks and minerals contained within their geographical limits.

It was soon found that the palæozoic rocks, as defined by Murchison, composed the greater part of the geological formations of America; that the carboniferous rocks, especially, were spread over a surface equal to England, France and Spain, united, and that the celebrated mines of the Ural Mountains were not only rivalled, but surpassed, by the copper mines of Lake Superior, the lead and iron ores of Wisconsin and Missouri, and, finally, by the golden *placers* of California and Oregon. As soon as the state geological reports were made public, several geologists wished to extend the geological map that Maclure had commenced in 1817; and notwithstanding the numberless difficulties inherent in such a work, in a country so vast and still so difficult to explore, in 1842 a *Geological Map of the Western States* appeared, by Byrem Lawrence. This work had not the publicity which it deserved, and has remained confined to a small number of persons capable of appreciating it.

A year later, in 1843, Mr. James Hall published in the fourth volume of the *Geology of New York*, a very interesting map, with many details, entitled *Geological Map of the Middle and Western States*. Then came Mr. Charles Lyell, who, in 1841 and 1842, collected all that had been published on America, made very judicious observations of his own on several parts of the tertiary rocks which are found in the Atlantic States, and published, in 1845, in his book entitled *Travels in North America*, a geological map of the United States and the British Provinces. This general map, which is the only one yet published, has rendered great service to the geology of this country, and has contributed more than any preceding work to our knowledge of the geological constitution of North America.

Notwithstanding these numerous works, and the excellent classifications and descriptions of the American palæozoic rocks, the parallelism of these strata with those of Europe was still uncertain, when M. de Verneuil, in the year 1847, visited this country, and succeeded in connecting them with the same rocks which Mr. Murchison and himself had classified in Germany, France, Scandinavia and Russia. M. de Verneuil published the results of his observations in a memoir inserted in the Bulletin of the Geological Society of France, which was translated and published in *Silliman's American Journal of Science.* The different groups of paleozoic rocks are now positively distinguished, and connected with the great geological epochs recognized in Europe.

In the geological map which accompanies this memoir I have adopted the divisions established by M. de Verneuil, and consequently Mr. Murchison's grouping of the palæozoic rocks. By adopting these divisions, the geological map of America can be compared with that of England, published by Murchison; with that of Germany, by Murchison, De Verneuil and D'Archiac; with that of Russia and Scandinavia, by Murchison and De Verneuil; and, finally, with that of Bohemia, published by Barrande. The advantage of comparison with so large a number of maps of other countries would have been a sufficient reason for adopting these divisions of the palæozoic rocks, even had I not also recognized the impossibility of exhibiting the numerous subdivisions introduced into these rocks by the geologists who have published the geological surveys of New York, Pennsylvania and Virginia. In the first place, the scale of my map, although one third larger than that of Mr. Lyell, is, nevertheless, too small to permit of a larger number of divisions; and then, what name shall be chosen for these subdivisions? Each geologist has given them a different name. For example, the *Upper Silurian* has been called in the State of New York *Oneida*, *Medina*, *Clinton*, *Niagara*, *Onon-*

daga, &c., *group;* in Pennsylvania and Virginia, *formation IV. V. and VI.*, and also *Levant series;* in Ohio, Indiana and Kentucky, the name is *cliff limestone;* in Tennessee, *gray limestone, or Harpeth and Tennessee river group;* in Iowa and Wisconsin, *formation III., b and c, of St. Peter's shell limestone*, or *upper magnesian limestone*, etc.

This single example shows the difficulty of selecting one name among these various classifications; and further, certain geologists have given too many local names to the subdivisions, and have assigned them an importance, in the American series, far too great, in the observation of so vast an extent of country as the United States and the British Provinces.

In a local study, as of the part or the whole of a State, the formations can and ought to be subdivided, to render the study of details as complete as possible. The case is not the same in a general study of a vast continent; here the details disappear in the masses, and one general character must mark each formation; without this precaution, you are overwhelmed in the details, which are as wearisome here as they are highly useful for a locality circumscribed and limited to a small field.

In publishing this geological map of America, with an explanatory text, my object has been to recapitulate, very concisely, the numerous observations of the geologists who have studied America, and to present, as completely as possible, the results which have been attained in the study of its geological formations. Perhaps this task is beyond my ability. Nevertheless, I have carefully studied all the published documents,* and even manuscript ones, which I have been able to procure, that relate to the subject; and I have travelled for three years in various parts of America, to verify the descriptions and geological maps that have been published. Many of the American geologists have assisted me in these researches, and I am especially grateful to Messrs. Richard Brown, of Sydney (Cape Breton), J. W. Dawson, of Pictou, Charles T. Jackson, of Boston, James Hall, of Albany, Benjamin Silliman, Jr., of New Haven, J. F. Frazer, of Philadelphia, and R. P. Yandell, of Louisville.

* See at the end of the memoir the titles of the geological works I have consulted, which present a bibliographic view of American geology.

CHAPTER I.

PALÆOZOIC ROCKS.

GENERAL SKETCH OF THE PHYSICAL FEATURES OF NORTH AMERICA.

BEFORE commencing a description of the geological structure of America, I wish to say a few words on the general form of this vast continent.

Bounded on the north by a series of islands and bays, whose number and form have not yet been wholly ascertained, notwithstanding the intrepid labors of Parry, Ross, Franklin, Richardson, Back, Rae, and others ; a line of hills, running from east to west, whose height does not seem to exceed twenty-five hundred feet, and whose composition is entirely granitic, separates the hydrographic basins of these arctic regions from the regions of the south ; the St. Lawrence, with its great lakes, washes the southern base of this system of mountains, which I shall call, after the example of the author of the *History of Canada*, Mr. F. X. Garneau, the *Lawrentine Mountains.*

The general system of the Alleghanies runs from north-east to south-west, and comes in contact with the Lawrentine system,— not, as is generally thought, in a direction which would reach Gaspé, but in a line which, passing by Newfoundland, would meet the coast of Labrador, if prolonged to the east. This Appalachian system, more elevated than the preceding,— for it attains four thousand, five thousand, and six thousand feet,— divides the hydrographic basin of the regions of the south into two parts, one of which includes the whole line of the Atlantic coast, extending from the River St. John, New Brunswick, to the Altamaha River, in Georgia ; and the other reaches from the sources of the Mississippi and Ohio to the Gulf of Mexico.

Finally, the system of the Rocky Mountains, mainly running from south to north, bounds the southern hydrographic basin on the west, and leads the waters of the Missouri, the Platte, the Arkansas, and the Rio Grande del Norte, into the Mexican gulf.

Independently of the three great mountain chains just pointed out, the whole of the country comprised in this immense triangle is an elevated plain, often intersected by deep ravines, which, after some undulations, gradually descends to the low and sandy coast, that forms the outline of the sea, from New York to Matamoras in Mexico. These undulations belong to two very different systems of hills:—The one that separates the waters of the basins of the Ohio and the Mississippi from the basins of Lake Michigan and Lake Erie is formed of small hills belonging to the sedimentary rocks, which compose all the surrounding country, and which have not been entirely carried away by the great denudation that swept the country at the time of the elevation of the Alleghanies. The other system of undulations is formed by three small masses of granite, that run parallel to each other from east to west, and were little islands in the primitive sea in which the palæozoic rocks were deposited. These primitive islets, of the same geological age as the Lawrentine Mountains, are found in the States of Missouri, Arkansas and Texas, called by geographers the Ozark Mountains, and represented as in one line; in reality, this representation is erroneous. They are in three parallel lines, and so distant as not to be properly considered as a chain of mountains, but as a system of hills, which I shall designate by the name of the *Ozark system.*

The fundamental rocks of the Lawrentine, Rocky, Alleghany and Ozark Mountains are composed of eruptive rocks, granitic, syenitic and porphyritic, forming the frame-work of plutonic origin which supports the basins of sedimentary rock that surround them. The first sedimentary beds deposited after the crust of the earth had become solidified were submitted to various metamorphic action by the often repeated injection of ignited matter in a liquid state, and also by the high temperature that still prevailed at the surface, which caused much more numerous chemical combinations than those that take place in our day. These first stratified rocks, thus modified, form gneiss, mica-schist, slate and marble,—which have often been classed as particular and exceptional formations, such as the *Cambrian system* of Mr. Sedgwick, in England, and the *Taconic system* of Mr. Emmons, in America. These systems ought to disappear from geological classification, for they give two names to the metamorphic rocks, of which they are integrant parts, in all regions where these beds have been observed. Not wishing to consider more at length the metamorphic rocks, before describing the beds in their normal condition, without alteration, I will begin by studying the bands of sedimentary formation that surround the groups of eruptive rocks, and which have not been influenced, in their lithological composition, by the action of heat.

On looking at the map, the reader will perceive a band of Silurian rock, which we have distinguished in many places by two different shades of the same color, to mark the two great divisions of Upper and Lower Silurian which compose this formation. The insufficiency of geological explorations, and the want of published documents, obliged us to omit this distinction in several portions of the map. It will require many years to supply this want; and, besides, it is sometimes impossible to unite the rocks of certain districts with these two great divisions, on account of the rarity of fossils; and the observer must be content with their synchronism with the general formation.

LOWER SILURIAN.

General Character.— The first strata are thick beds of very hard sandstone, rose-colored, or whitish-gray. Then comes a series of strata of compact limestone, blue, often blackish, with quite numerous fossils, the whole surmounted by schistose, slaty clay, of a deep black or blue-gray color.

Synonima.

State of New York.— *Potsdam sandstone, calciferous sandrock, Black River limestone group, Trenton limestone, Utica slate, Hudson River group.*— Messrs. Emmons and Hall.

Vermont.— *Isle La Motte limestone,* instead of *Black River limestone,*— the other divisions have the same names as in the State of New York.—Mr. Thompson.

Lower Canada (District of Gaspé and Quebec).— *Conglomerate limestone, Tourette's sandstone, Graptolite schist.*—Mr. Logan.

Upper Canada and District of Montreal (Lower Canada).—The same names of subdivisions as in the State of New York.

New Jersey, Pennsylvania and Virginia.—*Formation I., II. and III., or Primal and Matinal series.*— Messrs. W. B. and H. D. Rogers.

Ohio and Indiana.— *Blue limestone.*— Messrs. Locke and Owen.

Tennessee.— *Stone River group and Nashville group.* — Mr. Safford.

Wisconsin, Minnesota and Iowa.— *Formation* 1, *a*, *b*, *c*, *d*, *e*, *f*, or *lower sandstone; Formation* 2, *a*, *b*, *c*, or *lower magnesian limestone*, including *the Galena limestone; Formation* 3, *a*, or part of *St. Peter's shell limestone.*—Mr. D. D. Owen.

In most of the localities where the Lower Silurian strata are observed, they may be separated into three distinct divisions, which are found constantly with the same characteristics throughout the band of Silurian rocks which accompany the Lawrentine Mountains and the Alleghanies. The first of these divisions, which I shall call the *Potsdam formation*, is a very hard sandstone, of subcrystalline texture and very diffuse stratification. There are in it very few fossils, at least as to variety of species; the most characteristic are the *Lingula*, and the *Obolus* or *Ungulites*. The *Lingula antiqua* (Hall, Pl. I. fig. 1) is a species of Brachiopod Mollusk, which is often found in great numbers in these first strata of Silurian. It is found in New York, Michigan and Wisconsin. This sandstone is sometimes calcareous, and then it is distinguished from *Potsdam sandstone* by the special name of *calciferous sandrock.* Its thickness varies with the different localities in which it is found, and depends on the more or less horizontal position of the bed; nevertheless it may be said to vary from five hundred to two thousand feet.

These Lower Silurian strata of America have many analogies with the same Silurian strata in Russia and Scandinavia, where the *Ungulite grit* of the environs of St. Petersburg presents the same structure, the same fossils, and the same stratigraphic position. Of great extent in Canada, at Montmorency Falls; in the State of New York, at Potsdam; in New Jersey, Pennsylvania, Virginia, North Carolina and Georgia; it has not been discovered in Ohio, nor Tennessee, where, however, the two other divisions of Lower Silurian are found. Mr. D. D. Owen has shown it to be of a considerable depth in the west of Wisconsin, where it presents a larger number of *Lingula* and *Orbicula* than in any other part of America. Mr. Owen describes it under the name of *lower sandstone of the Upper Mississippi.* In Canada this sandstone is observed to prevail on the borders of the Ottawa, of the Two Mountains River, of the St. Lawrence (Thousand Islands), near Lake Simcoe; on the Escanaba River, near Lake Michigan, and at Janesville, near Milwaukie. Several geologists regard the sandstone of Lake Superior as belonging to this Lower Silurian; but we think they are in error, as we shall endeavor to demonstrate in describing the *New Red Sandstone* and the Systems of Mountains.

The second division is composed chiefly of blue limestones, with intercalation of clay of the same color. These limestones have been divided into two parts by the authors of the *Geological Survey of New York*, who have called the lower part *Black River group*, and the upper *Trenton limestone.* For the sake of conciseness, I shall reckon these subdivisions under the second division of the Lower Silurian, or *Trenton formation.* It is in this division that for the first time is presented a complete *fauna*, representing the first degree of the biologic development of our planet. The first division of the Lower Silurian offers only a few species of animals, rarely to be found, and mostly in a bad state of preservation; while this second division shows at once an immense development of crustacea, mollusks and polyps. Mr. James Hall describes two hundred and ninety-five species belonging to this division, in his valuable work, *The Paleontology of New York.* This number will certainly be increased by researches made in other parts of America, and even in the State of New York itself. We will cite here only the most characteristic species, and those which are found in the same geognostic horizon in Europe. Such are,

Illænus crassicauda, Wahl. (Pl. I. fig. 2.)—A kind of Trilobite which is found in Canada, New York, Vermont, Pennsylvania, Virginia and Tennessee, and occupies the same relative place in England, France, Sweden and Russia.

Orthoceratites communis, Wahl. (Pl. I. fig. 3.)—One of the most characteristic species of cephalopods of the Lower Silurian beds of Russia and Sweden. This species is common in New York, Pennsylvania, Canada, the Mingan Islands, on the coast of Labrador, and in Newfoundland, near the Straits of Belle Isle.

Bellerophon bilobatus, Murch. (Pl. I. fig. 4.)—A species of

Gasteropod, very abundant at Trenton Falls and in many other parts of the State of New York; it is found also in Canada, Ohio, Wisconsin, in England, and at Christiana, Norway.

Spirifer lynx, Eichw. (Pl. I. fig. 5 and 5 a.) — Is a Brachiopod Mollusk, one of the most common in America, as well as in Europe; it is found from Newfoundland to Tennessee, and on the Upper Mississippi.

The Orthis testudinaria and *Verneuili*, Dalm. (Pl. I. fig. 6 and 6 a, fig. 7.) — Species of Brachiopods, related to the Spirifer, are two fossils which are found equally in Europe and America.

Lastly, the *Chætetes petropolitanus*, Pand. (Pl. I. fig. 8 and 8 a.) — Is one of the most characteristic and most numerous polyps of the second division of the Lower Silurian. It is found from Newfoundland to Tennessee and Alabama; in Arkansas and Minnesota. In Europe it is very common in the environs of St. Petersburgh and of Christiana (Norway).

This second division of Lower Silurian is more easily recognized than are the two others, on account of the large number of fossils it contains, and the considerable depth it presents in some places, — it being from four to five thousand feet in Pennsylvania. It has been indicated in a great many points in the United States and the British Provinces. Recognized by Bayfield at the Mingan Islands and in Newfoundland, it pursues the whole length of the St. Lawrence, then the River Richelieu, Lake Champlain and the Mohawk, ascends the Ottawa nearly to Lake Nipissing, follows the east and north coasts of Lake Ontario, enters Lake Huron by Georgian Bay, is found at Lake Michigan in Green Bay, and continues to Wisconsin and Illinois, where it contains the rich lead mines of Galena. Finally, it enters Minnesota, where it forms part of the descent for the falls of the Mississippi, at St. Anthony. Its existence has been verified on Lake Winnipeg, near Fort Alexander, along the Red River, at Lakes Abbitibbe and St. John, in the Hudson's Bay Territory.

Following on the south the line of the Alleghanies, it forms, from New Jersey to the extreme north-east point of Alabama, a series of the highest longitudinal mountains and valleys of the Appalachian chain. Most of the *blue limestone* of the environs of Cincinnati connects itself with this division, as well as the *Stone River* and *Nashville groups* of Tennessee. Mr. Rœmer has verified Silurian strata which appear to belong to this second division, near the Rio San Saba, in Texas. And the environs of Hot Spring, in Arkansas, as well as the *Galena limestone*, found near Potosi, in Missouri, seem to contain strata which date from the same epoch.

A third division, composed of argillaceous schist, very fissile and resembling slate, terminates the Lower Silurian. It is distinguished in the State of New York and in Canada by the names *Utica slate* and *Hudson River group*, and occupies the first rank in the series of sedimentary rocks, owing to the great thickness of its beds and the extent of country which it covers. Fossils are rare in this division; the only ones are Graptolites, sometimes in great abundance, and fragments of trilobites, especially the *Trinucleus Caractaci.**

The recognition and the study of these beds are rendered difficult and monotonous by the scarcity of fossils. The basins of the lower St. Lawrence and of the Hudson are regions especially belonging to this division, which extends from near Cape Rozière, at the extremity of Gaspé, to Virginia, where it still has a depth of three thousand feet; thus it may be said to form a band of rock which, varying from three to ten miles in width, extends uninterruptedly over seven degrees of latitude, without important variation either in its mineralogical constitution or its stratification. This division is found again in Upper Canada, and at the Bay des Noquets, in Lake Michigan. In Wisconsin and Minnesota it is perhaps represented by the first subdivision of the formation 3, or *St. Peter's shell limestone* of Mr. Owen; and in Ohio, Kentucky and Tennessee, the upper part of the *blue limestone* and of the *Nashville group* may also be regarded as equivalent to it; but these are only suppositions, and we are inclined to look upon this third division of Lower Silurian as mostly confined to the eastern and northern regions of America.

We have verified it at Lake Superior, where it forms the Cataract of Kakabeka, on the River Kaministiquia. Russia and Scandinavia, countries that have so much resemblance to North America in their

* *Trinucleus Caractaci*, Murch. (Pl. I. fig. 9.) This species of trilobite is very characteristic of the third division of Lower Silurian. It is abundant in Lower Canada, and in the States of New York, Pennsylvania, Kentucky, Ohio and Tennessee. In Europe it is found at the same geological level, and especially in the Caradoc sandstone of England.

Graptolites pristis, Hall. (Pl. I. fig. 10.)—Species of Bryozoa, according to Agassiz very common both in America and Europe.

chief geological features, do not show this third division, unless the black schist that forms the top of Kinnekulle Hill, between Gottenburg and Stockholm, may be related to it, which is uncertain; while the environs of St. Petersburgh and Norway present the second division, with as fine a development, perhaps, as America, both for the fauna and the depth of the beds. Mr. Murchison, in his learned work on Russia, distinguishes this second part of the Lower Silurian by the name of *Pleta* or *Orthoceratites limestone.*

UPPER SILURIAN.

General Character.— Light-gray limestone, becoming sometimes blue-gray, with interposition in many places of bluish-gray clay.

Synonima.

State of New York.— *Gray sandstone*, *Oneida conglomerate*, *Medina sandstone*, *Clinton group*, *Niagara group*, *Onondaga salt group*, *Water-lime group*, *Pentamerus limestone*, *Delthyris shaly limestone*, *Encrinal limestone*, *Upper Pentamerus limestone.*— Messrs. Hall and Vanuxem.

Lower Canada.— *Limestone and schist of Gaspé* (chiefly). — Mr. Logan.

Upper Canada.— The same names as the State of New York.

Pennsylvania and Virginia.— *Formation IV.*, *V.*, *VI.*, or *Levant series.*— Messrs. Rogers.

Ohio, Indiana and Kentucky.— *Cliff limestone* (chiefly).— Messrs. Locke, Owen and Yandell.

Tennessee.— *Gray limestone*, or *Harpeth and Tennessse River group* (chiefly).— Mr. Safford.

Minnesota and Iowa.— *Formation* 3, *b and c of St. Peter's shell limestone*, or *upper magnesian limestone.*— Mr. Owen.

Wisconsin.— *Waukesha limestone and the geodiferous limestone* of M. Eaton.— Mr. Lapham.

The Upper Silurian, though generally following the direction marked out for it by the Lower Silurian, differs a little in its geographical distribution; thus it is not found on the borders of the St. Lawrence.

Beginning near the Straits of Belle Isle, Newfoundland, it

forms the whole Island of Anticosti, and part of Cape Rozière and the point of Gaspé, and extends to the south of the mountains of Notre Dame, from whence it crosses the Matapediac River, gains the Madawaska and the Temiscouata Lake, ascends the St. John, crosses the Chaudière and St. Francis, and finally reaches Lake Memphramagog, on the borders of Canada and Vermont, penetrates a little way into this state, and loses itself in the ramifications of the Green Mountains. The Upper Silurian is found again at several points in New Brunswick, and at Eastport, in Maine. Mr. Logan has verified it on Lake Temiscaming, in the Hudson's Bay Territory. Beginning in the north-eastern part of Alabama, the Upper Silurian beds may be followed, without interruption, through the whole chain of the Alleghanies, to the State of New York, where it has a larger development, especially on the southern border of Lake Ontario. Messrs. Hall and Vanuxem have distinguished a large number of clearly-marked subdivisions in this part of America. Crossing Upper Canada, it forms a part of the Manitoulin Islands, in Lake Huron, and the north and west sides of Lake Michigan; then, forming the upper part of the State of Illinois, it ascends into Iowa and Minnesota, where it forms the upper part of the falls of the Mississippi, at Fort Snelling. The environs of Cincinnati, Louisville and Nashville, also present the Upper Silurian; it forms the cliffs of the numerous hills which are found in this part of the basins of the Ohio and the Tennessee, and which, for this reason, has been called, by Western geologists, *cliff limestone.*

Several geologists, as well as officers of the United States army, have found Silurian beds in diverse regions, which it is not possible to connect with certainty with either of the two formations of this group. Messrs. Dawson and Brown have recognized the Silurian at Cape Breton and in Nova Scotia; Mr. Jackson, in several parts of Maine; Mr. Featherstonhaugh, in Missouri and near Little Rock, in Arkansas; Mr. Rœmer, in Texas; lastly, Messrs. Frémont, Stansbury and Wislizenus, have found, at several points in the Rocky Mountains and in New Mexico, rocks which belong, without doubt, to the Silurian. The fossils which we have seen to be so rare in the third division of Lower Silurian show themselves in great abundance in the Upper Silurian. The species are different from those of the Lower Silurian, though they are analogous to them, and belong to the same genera.

We shall only cite the most characteristic, and those which are common to Europe and America.

Orthoceratites annulatus, Sow. (Pl. II. fig. 1.) — A species of Cephalopod Mollusk, which is abundant in England, in the Island of Gothland (Sweden), as well as in the States of New York and Pennsylvania, at Madawaska in Maine, and in Tennessee.

Terebratula cuneata, Dalm. (Pl. II. fig. 2.) — A species of Brachiopod, very characteristic and well defined, is found at Lockport. New York, in England, in the Island of Gothland, and in Bohemia.

Pentamerus oblongus, Sow. (Pl. II. fig. 3.) — This species of Brachiopod, very common in England and in Bohemia, is found almost universally in the Upper Silurian of the United States; thus it is found at Anticosti Island, at Gaspé, in Maine, New York, Pennsylvania, Virginia, North Carolina, Ohio, Indiana, Tennessee and Kentucky; in Upper Canada, on Lake Huron, in Wisconsin and Minnesota.

Spirifer sulcatus, Dalm. (Pl. II. fig. 4.) — A species of Brachiopod, found in the Island of Gothland, and at Lockport, New York.

Spirifer crispus, Dalm. (Pl. II. fig. 5.) — A species related to the preceding, and which is found in the same localities in Europe and America.

Orthis hybrida, Sow. (Pl. II. fig. 6 and 6 a.) — A species of Brachiopod, very common in many parts of the State of New York, Canada and Tennessee. It is also found in England, and in the Island of Gothland.

Leptæna depressa, Sow. (Pl. II. fig. 7.) — This species of Brachiopod has a smaller variety, which is found in the Lower Silurian. But the true *Leptæna depressa*, which is that of our figure, is found at Dudley, England; in the Island of Gothland; in Bohemia; at Lockport and Rochester, State of New York; in Tennessee, Illinois, Minnesota; at Eastport, in Maine; at Gaspé, and at Anticosti, Lower Canada.

Hypanthocrinites decorus, Phil. (Pl. II. fig. 8.) — This remarkable Crinoid, which is found in England and Sweden, is frequently met with in the environs of Lockport, New York, and of Perrysville, Tennessee.

Favosites gothlandica, Goldf. (Pl. II. fig. 9.) This species of polyp is found in America, as in Europe, in the Upper Silurian and the Devonian. It is met with in abundance in the Island of Anticosti; at Madawaska, Maine; Lockport and Schoharie, New York; Chicago, Illinois, and in Minnesota and Kentucky.

Catenipora escharoides, Lamk. (Pl. II. fig. 10.) — A very common and very characteristic species of polyp, in the whole region of Upper Silurian, both in Europe and America.

The Upper Silurian in America often present beds of rock-salt; and the salt works of Syracuse, in the State of New York, are built on this formation. I will mention, also, in passing, the Cataract of Niagara, which is entirely formed of rocks belonging to the Upper Silurian.

DEVONIAN FORMATION.

General Character. — The first strata of the Devonian are formed of whitish-gray limestone, containing a great number of fossils. Then numerous beds of black schistose clay are superposed, as in the States of New York and Pennsylvania; and finally, in some places, as Gaspé and Catskill, these beds are crowned by very thick beds of red sandstone, with very few fossils.

Synonima.

State of New York. — *Oriskany sandstone*, *Cauda galli grit*, *Schoharie grit*, *Onondaga limestone*, *Corniferous limestone*, *Marcellus slate*, *Hamilton group*, *Tully limestone*, *Genesee slate*, *Portage group*, *Chemung group*, *Old Red Sandstone*. — Messrs. Hall and Vanuxem.

Lower Canada. — *Calcareous schist of Gaspé* (the upper part), *and the sandstone of Gaspé*. — Mr. Logan.

Upper Canada. — *Upper limestone*. — Mr. Murray.

Pennsylvania and Virginia. — *Formation VII.*, *VIII.*, *IX. and X.*, or *Premedidial*,* *Medidial*, *Postmedidial* and *Ponent series*. — Messrs. Rogers.

* These terms, *matinal*, *levant*, *premedidial*, *medidial*, &c., drawn from the divisions of the day, are entirely different from common geological classifications; and, although proposed by the Messrs. Rogers at a meeting of the Association of American Geologists, in 1844, this method has only been adopted by its authors.

Ohio and Indiana. — The upper beds of *cliff limestone*. — Messrs. Locke and Owen.

Kentucky. — *Ohio Falls group*, divided by Mr. Clapp into *upper and lower coralline beds; 2d, middle or shell bed;* and 3*d*, *upper or limestone bed*. — Messrs. Yandell and Shumard.

Tennessee. — The upper part of *gray limestone*. — M. Safford.

Wisconsin and Iowa. — *Formation of the Red Cedar River*. — Mr. Owen.

Michigan. — *Mackinaw limestone*. — M. Houghton.

The Devonian formation, although it occupies an important place in the series of American rocks, does not present the same character of constancy and homogeneity, in its geognostic constitution and its geographical distribution, that we have seen in the Upper and Lower formations of Silurian.

In the States of New York and Pennsylvania the Devonian is composed of very numerous and very deep strata; the whole formation attains a depth of fourteen thousand feet; while in the States of Indiana, Kentucky and Tennessee, the depth is only from one to two hundred feet. In the eastern part of Canada, at Gaspé, which is the most northerly point where this formation has been observed in America, the development of strata is also very considerable, and approaches the immense depth observed in the State of New York. Very fossiliferous sandstones form the first Devonian beds in Pennsylvania and New York; then comes a great extent of marl and clay, presenting in certain localities quite numerous fossils; and lastly, the whole is crowned by very deep red sandstone, especially at the Catskill Mountains, New York, at the base of the Alleghany Mountains, Pennsylvania, and at Gaspé, Lower Canada. This red sandstone represents, without doubt, the Old Red Sandstone formation of Wales and Scotland; it has the same geognostic and paleontologic character; for in America, as in Europe, nothing is found in it but the remains of fishes. So that this upper part of the Devonian must be regarded as a local group, which is found at several points of the Devonian basins of Europe and America.

In the basins of the Ohio and the Mississippi, and on Lakes Erie, Huron and Michigan, the Devonian is composed of one group of strata, containing very fossiliferous limestone beds, of a light-gray color, often whitish, and following the limestone strata of the Silurian, with which they have much lithologic analogy. Since Mr. Agassiz has recognized carboniferous fishes and goniatites in the *black slate* of Ohio, this group ought to be placed in the Lower Carboniferous, of which it forms the base. The Island of Mackinaw, the Ohio Falls near Louisville, and Perry County in Tennessee, have become classic points for the American Devonian, on account of the great number of fossils which are found there, and their identity in species, for the most part, with those found in the Eifel, in the Hartz Mountains, and in Devonshire.

The Devonian is placed in two very distinct basins,— one east and the other west of the Green Mountains,—whose elevation took place at the end of the Silurian period, and previous to the formation of this deposit. Recognized at Gaspé, on the Rivers Ristigouche and St. John, it forms in this region the border of the coal basin of New Brunswick; but it has not yet been recognized in Nova Scotia, Cape Breton or Newfoundland. West of the Green Mountains and the Hudson River, the Devonian spreads over a vast extent of country, forming a part of the Alleghany chain from Schoharie and Catskill to the south-east of the city of Tuscaloosa, Alabama; to the west it extends through the southern part of the State of New York, forms the whole contour of Lakes Erie and St. Clair, a part of the peninsula of Michigan, the Island of Mackinaw and the southern side of Lake Michigan; then it reaches the north of Indiana and Illinois, and enters Iowa by the valley of the Red Cedar River; from thence it extends probably towards the Missouri, where it disappears under the beds of Cretaceous rocks. The Silurian valleys of Cincinnati and Nashville are surrounded by this formation, which extends by the Tennessee valley towards the Mississippi, where it forms a narrow band round the primitive islet, which is on the south of the State of Missouri.

The Devonian has not been verified with certainty in Arkansas, Texas, or New Mexico; but it is probably to be found there also. Captain Stansbury recognized it on the western side of the Rocky Mountains, not far from Fort Laramie. The most characteristic fossils, which are common to the Devonian basins of Europe and America, are:

Calymene bufo, Green. (Pl. III. fig. 1.) — This Trilobite is found everywhere in the Devonian beds, from Altaï, Russia, to Tennessee and Iowa; that is, over a fifth part of the surface of the globe;

very common in the Eifel, it is equally so at Schoharie and Moscow, New York; at the Ohio Fall, Kentucky; in Virginia, Tennessee, Indiana, and at Gaspé, Lower Canada.

Terebratula reticularis, Linn. (Pl. III. fig. 2, 2 a.) — This species presents two varieties. The smallest, which is found in the Upper Silurian, as well in America as Europe; and the other, nearly twice the size, which is exclusively Devonian. It is the last variety which we present in the plate; it is very common in all the regions where the lower part of the Devonian is observed, both in America and Europe, and may be regarded as one of its most characteristic fossils.

Terebratula concentrica, Von Buch. (Pl. III. fig. 3.) — In America, as in Europe, this species of Brachiopod is abundant in the lower and middle beds of the Devonian. The environs of Louisville, and many localities in the State of New York, present it in great abundance.

Pentamerus galeatus, Dalm. (Pl. III. fig. 4, 4 a.) — This species, common to the Upper Silurian and the Devonian of Europe, is in the same geological position in America; but on this side of the Atlantic it is especially found in the Devonian division. It is most abundant in the States of New York and Tennessee.

Spirifer mucronatus, Conr. (Pl. III. fig. 5, 5 a.) — This Brachiopod is very common in the State of New York, as well as in Ohio and Tennessee. Although much larger than the species of the Eifel, it is identical with it.

Spirifer Verneuili, Murch. (Pl. III. fig. 6.) — This species is common in France and Belgium, and is equally so in the State of New York.

Spirifer heteroclitus, Def. (Pl. III. fig. 7, 7 a.) — This species, much more common than the last, is found in a much greater number of localities. In the Timans Mountains, in the north of Russia, at the Eifel, in Brittany, in Spain and in England. In the United States it is very frequent at the Ohio Fall, and at Charlestown Road, Indiana; in Ohio, New York, Pennsylvania and Tennessee.

Chonetes nana, De Vern. (Pl. III. fig. 8, 8 a, 8 b.) — This small species of Brachiopod has only been found in Russia, and in the environs of Louisville, Kentucky.

Zaphrentis gigantea, Rafin. (Pl. III. fig. 9.) — This fossil, first found by Lesueur, and described by Rafinesque, is one of the most characteristic of the American Devonian. It is found in great abundance at the Ohio Fall, where it forms nearly the whole of several beds of limestone; it is also very common at the Island of Mackinaw.

CARBONIFEROUS GROUP.

We have now reached a series of rocks whose strata contain the most useful minerals, and the materials most necessary to supply the wants of modern industry. Here are to be found coal, iron, the quartzose sandstone for the glass manufacturers, and in America they contain, besides, vast beds of gypsum and rock salt. Such great mineral riches, disposed, so to speak, in beds which succeed each other and are arranged like the leaves of the same book, give to the States which are happy enough to contain a portion of the Carboniferous group an inestimable commercial and industrial value, and will one day afford them a decided superiority over the neighboring country.

A glance at our map will show what an immense extent is embraced by this formation. No part of the known world offers so great a development of Carboniferous rocks. From Newfoundland to Vancouver's Land, that is, through the whole width from east to west of North America, a road might be traced almost wholly upon the rocks that belong to the Carboniferous, with two little interruptions, one in the New England States, and another in the Rocky Mountains. In truth, so great an extension of the Carboniferous region may well astonish the boldest spirit of geological speculation; and one can hardly yet imagine all the changes that will be brought about in the wealth and modes of life of the American world by the immense influence to be exerted on society through this abundance of the materials which have become the keystone to the industrial edifice of the nineteenth century.

Two shades of the same color on the map mark the two great divisions of the Carboniferous, as before of the Silurian group. Here, again, as in the Silurian, this distinction could not always be preserved; sometimes owing to the small scale of the map, and sometimes from the want of sufficiently accurate knowledge.

LOWER CARBONIFEROUS.

General Character.— This formation is especially characterized, first, at the base, by bituminous black schist, or black slate; then by an immense development of fine-grained sandstone or psammite and of whitish-gray limestone, containing a great number of fossils. In several regions, as Nova Scotia, Cape Breton, Newfoundland, Ohio, Osage country and Great Salt Lake, they contain gypsum and rock salt.

Synonima.

Nova Scotia, Cape Breton and New Brunswick.— *Mountain limestone*, or *gypsiferous formation* of Messrs. Dawson and Brown. *Limestone, red sandstone, gypseous marl and gypsum of the New Red Sandstone* of Mr. Gesner.

Pennsylvania and Virginia.— *Formation XI. and XII.*, or *Vespertine series.*— Messrs. Rogers.

Ohio.— *Bituminous slate, Waverley sandstone series*, or *fine-grained sandstone and conglomerate.*— Messrs. Briggs and Locke.

Indiana and Illinois.— *Bituminous slate, fine-grained sandstone of the knobs and oölitic limestone series.*— Mr. Owen.

Kentucky.— *Slate beds, Waverley sandstone, with the encrinital limestone of the button-mould knobs, Carboniferous limestone of Mammoth Cave.*— Messrs. Yandell and Shumard.

Tennessee.— *The black slate, siliceous group, encrinital limestone, oölitic limestone.*— Messrs. Troost and Safford.

Missouri, Iowa, Nebraska and Great Salt Lake.— *Carboniferous*, or *Mountain limestone.*— Messrs. Nicolet, Owen, King and Stansbury.

M. de Verneuil, in his excellent memoir on *the Parallelism of the Palæozoic Formations of North America with those of Europe*, says that "of all the principal divisions which form the Palæozoic rocks of America, the Carboniferous group is that which is the most clearly characterized, and which has the most in common with the European deposits of the same epoch."

In truth, the rocks and fossils are very similar, and often identical. In the Lower Carboniferous, which corresponds to the *Carboniferous or Mountain Limestone* of England, there are no beds of coal; but the beds of argillaceous schist, sandstone and limestone, which compose it, present often, associated with them, thick beds of gypsum and rock salt, wholly analogous to the gypsum and rock salt of the Trias of France and Germany, which has caused several American geologists to connect them with the New Red Sandstone.

In Nova Scotia, where this rock has been studied in much detail by Messrs. Dawson and Brown, it is found to be composed especially of reddish sandstone, with clay of the same color, conglomerate, and porous whitish limestone, often very fossiliferous (Windsor), and also in many places where this formation is exposed it shows thick beds of gypsum. Its whole thickness is six thousand feet, and even more in some places.

In Pennsylvania, Maryland and Virginia, the Lower Carboniferous is chiefly represented by red schist and siliceous conglomerate, whose thickness is not more than three thousand feet. Beyond the Alleghanies this rock occurs again under the coal beds, and extends over the greater part of the Western States of the Union. This formation is composed of black slate, very fine sandstone, of a whitish color, and limestone, often siliceous and oölitic, of the same color as the sandstone; and contains also in several places mines of very rich iron ore (Tennessee), salt springs (Ohio), and gypsum in great quantities on the borders of the Canadian River, not far from old Fort Holmes, west of Fort Gibson. Mr. Nicolet has recognized this formation in the environs of St. Louis, on the River Des Moines, and also in ascending the Missouri as far as Council Bluffs, where it is covered by the strata of the Cretaceous group. Capt. Stansbury, in his *Expedition to the Great Salt Lake*, also recognized the Lower Carboniferous in the environs of Fort Laramie, at the foot of the Wind Mountains, and on the western shore of the Great Salt Lake. Lastly, Mr. Dana, in the *Geology of the Exploring Expedition of the United States*, recognizes it also at Puget's Sound and Vancouver's Land, north of Oregon. So great an extension of the *Mountain Limestone* of England is one of the most remarkable geological facts, as it forms the finest geological horizon that exists upon our planet; especially if it be borne in mind that this formation is found in Russia, the Altaï Mountains, in Germany, France, Belgium, Spain, England, and even in Australia, and at Lake Titicaca, in South America.

The list of characteristic fossils, common to Europe and America, which follows, is distinguished from the preceding ones by the fact that all these fossils have been also found in Russia by M. de Verneuil, who has described and drawn them in the second volume of the *Geology of Russia in Europe and the Ural Mountains.* It is also noticeable that the fauna of the Lower Carboniferous is very rich in *Productus*, a species of Brachiopod, allied to the Spirifer and the Terebratulæ.

The fossils characteristic of the Lower Carboniferous are:

Nautilus tuberculatus, Sow. (Pl. IV. fig. 1.) — A Cephalopod Mollusk; is very common in Russia, Yorkshire, Ireland, Belgium, and also in Nova Scotia, New Brunswick, Ohio, Pennsylvania and Kentucky.

Terebratula Roissyi, Lév. (Pl. IV. fig. 2, 2 a.) — This species, common in Europe, is found at Windsor, Nova Scotia, Cape Breton, and Newfoundland, in Virginia, near Louisville, Kentucky, in the State of Ohio, and upon the Wabash River, Illinois.

Spirifer lineatus, Mart. (Pl. IV. fig. 3, 3 a.) — This species, very characteristic of the Mountain Limestone in England, Belgium, France and Russia, is found almost throughout this formation in America, from Newfoundland to Fort Laramie, Rocky Mountains.

Spirifer striatus, Mart. (Pl. IV. fig. 4, 4 a.) — A species common at Louisville, Glasgow (Kentucky), in Tennessee, Missouri, Iowa, Arkansas, Texas, and Nova Scotia. Russia, and many other countries in Europe, also present it in abundance.

Orthis crenistria, Phill. (Pl. IV. fig. 5, 5 a.) — This Brachiopod passes also, sometimes, into the Devonian rocks of America and Europe, but it is much more abundant in the Lower Carboniferous, and there it is larger. It is found in Pennsylvania, Ohio, Indiana, Kentucky, Tennessee, Missouri and Iowa.

Orthis Michelini, Lév. (Pl. IV. fig. 6, 6 a.) — Characteristic and very common in Europe. This species is also abundant in America, in the same regions as the preceding species.

Chonetes sarcinulata, Schlot. (Pl. IV. fig. 7.) — This species, very common in Russia, and abundant in France and Belgium, is found in Ohio and Kentucky, but is more rare than in Europe.

Productus semireticulatus, Mart. (Pl. IV. fig. 8.) — This fossil is certainly one of the small number that have enjoyed the privilege of spreading themselves over the greater part of our globe. It is found in the Carboniferous basin of the Gulf of St. Lawrence, and in those of Ohio, Tennessee, Alabama, Mississippi, Missouri, the Rocky Mountains, the Great Salt Lake, and Vancouver's Land. Europe presents it to us in almost all the parts where the Carboniferous group appears. Still more it is found in the Altaï (Asia), and at Lake Titicaca, in Bolivia, South America.

Productus Cora, D'Orb. (Pl. IV. fig. 9.) — This species of Brachiopod is as widely spread as the preceding. It is found in the United States, Bolivia, Spitzbergen, in Spain, Russia, and in the upper Punjaub (East India).

Productus Flemingi, De Vern. (Pl. IV. fig. 10.) — This Productus, like the two preceding, is wholly confined to the Carboniferous group, and is abundant in this formation from the Ural and Himalaya Mountains, in Asia, to Bolivia, in South America.

Productus punctatus, Sow. (Pl. IV. fig. 11.) — This species is the most common in North America,—it is almost universal in the Carboniferous. Capt. Stansbury has found it at the Great Salt Lake.

Fusulina cylindrica, Fisch. (Pl. IV. fig. 12, 12 a.) — This Foraminifer is found in abundance on the borders of the Ohio, in a limestone perforated throughout with little cavities left by the fusulinas, exactly as M. de Verneuil has described it in Russia on the borders of the Volga.

UPPER CARBONIFEROUS OR COAL MEASURES.

Above the carboniferous limestone is found sandstone, sometimes accompanied by conglomerate and the schists, which encloses the true coal. This formation is hardly to be distinguished from that of Europe, excepting that it covers a vast extent of country, and is of greater thickness. It also contains between the strata limestone of marine origin, which gives to it a mixed character, that it has not in Europe. Otherwise the coal, the fossil plants and fishes, which are found in the mines of America, present the same species and the same varieties as in England and France. The thickness of the coal rock of Nova Scotia — at least, as it has been approximatively stated — is more than ten thousand feet; in Pennsylvania, Ohio and Alabama, it attains seven or eight thousand feet.

Distributed over nearly half the coast of the Gulf of St. Lawrence, from St. George's Bay, Newfoundland, to Bathurst, in the Bay of Chaleurs, New Brunswick, the beds of coal are often seen exposed, even in cliffs that surmount the sea, seeming to hold themselves in readiness, so to speak, for the ships that cross the gulf. The mines of Sidney and Pictou are celebrated even in the United States; and their coal competes with that of Pennsylvania, Maryland and Virginia.

There is a small basin of anthracite coal at the south of the State of Massachusetts, and in the eastern part of the State of Rhode Island; which, when the coal deposits were made, was certainly connected with the basins of Nova Scotia and Pennsylvania, and seems to remain, in order to show the link which formerly united the beds of Pottsville to those of Pictou. It is especially remarkable for its geographical position and the metamorphic phenomena to which it has been subjected at the epoch of the granitic and porphyritic eruptions that gave birth to the chain of the Alleghanies.

After having crossed the *Blue Ridge*, and the first chain of the Alleghanies, a series of small coal basins is found placed in the valleys, and forming an anthracite region of great industrial value, because of its proximity to and means of communication with the largest cities of the United States, namely, Philadelphia, Baltimore, New York and Boston. These small detached basins, enclosed in the middle region of the Appalachian Mountains, are only, so to speak, an advanced guard of the great Alleghany bituminous coal-field, which forms the last and highest chain of these mountains, and extends to the east and to the south, from the boundary of the State of New York to the centre of Alabama.

In looking on the map, it will be easily seen that this coal-basin should join those of Illinois, Michigan, Missouri and Arkansas; and that the interruptions to these coal-beds are due to immense denudations, it may be at the time of the dislocation of the Alleghanies; or, it may have been at a later period.

Coal-beds have been found in Texas, near the Rio San Saba, in the Rocky Mountains, near Rock Independence, and in Vancouver's Land, and Puget's Sound, Oregon.

Future researches will discover new localities of this precious mineral in that immense region of the West, the Rocky Mountains and

Upper California, a country so little explored, and so little known even in a geographical point of view. Each year sees new expeditions, commanded by the officers of the corps of Topographical Engineers of the United States Army, and directed by the learned Colonel J. J. Abert, of Washington, travel over these immense solitudes, making topographical surveys, almost always accompanied by excellent notices on the geology, zoology and botany, of the regions visited; and in this way each year increases our knowledge of this part of the New World.

The number of fossil plants common to America and Europe is very considerable. As we can cite but a few, they will be the five most characteristic species, which are:

Calamites Cistii, Brongn. (Pl. V. fig. 1.) — This species of reed often attains great size. It is abundant in Pennsylvania, at Wilkesbarre and Carbondale, in Ohio and Indiana, and at Sidney, Cape Breton. In Europe it is found in France, England, and in Rhenish Prussia.

Neuropteris angustifolia, Brongn. (Pl. V. fig. 2.) — Fossils of this species are often found in the mines of Pottsville, Pennsylvania; at Sidney, Cape Breton, and Mansfield, Massachusetts.

Neuropteris Loshii, Brongn. (Pl. V. fig. 3.) — This species is one of the most common, both in Europe and America.

Sigillaria Sillimani, Brongn. (Pl. V. fig. 4.) — This plant, frequent in the coal-basin of Saarbruch, Prussia, is also abundant at Carbondale and Wilkesbarre, Pennsylvania; at Joggins Mines, Nova Scotia, and at North Sydney, Cape Breton.

Lepidodendron elegans, Brongn. (Pl. V. fig. 5.) — The mines of Sidney, Cape Breton, contain an innumerable quantity; it is also found in almost all the coal-mines of the New and Old Continent.

CHAPTER II.

SECONDARY ROCKS.

NEW RED SANDSTONE FORMATION.

THIS formation of sandstone is spread over a large part of America; but it never attains great width, and is much more remarkable for its geographical distribution, and the minerals which have accompanied its formation, than for the development of its strata and its horizontal extension.

The first geologists who studied America confounded this formation with the Old Red Sandstone of Europe, and its relative age was fixed only after a long controversy. Several of the persons employed by the governments of the United States and Canada even now continue to regard it as belonging to the epoch of the ***Potsdam sandstone;*** that is, to the first division of Lower Silurian. This last opinion is sustained by Messrs. Logan, Foster and Owen, who regard the sandstone of Lake Superior as the same with the Potsdam; but in our opinion it is New Red Sandstone, similar in all points to the New Red Sandstone which has been positively recognized in other parts of America.

The general character of the rocks composing this formation is a development of red sandstone, sometimes whitish-gray, in thin and often schistose strata, of variable hardness, though generally very tender, and having the lithologic form well known in this country by the name of *freestone.*

Synonims.

Magdalen Islands. — *New Red Sandstone.* — Mr. Baddeley.

Nova Scotia. — *New Red Sandstone.* — Messrs. Jackson and Alger. *New Red Sandstone* of Minas Basin. — Mr. Dawson.

Prince Edward's Island. — *Coal measures.* — Messrs. Henwood and Lyell.

Gaspé, Bay of Chaleurs. — *Conglomerate limestone and red sandstone.* — Mr. Logan.

New Brunswick. — *New Red Sandstone* (in part). — Mr. Gesner.

Massachusetts. — *New Red Sandstone.* — Mr. Hitchcock.

Connecticut. — *New Red Sandstone.* — Mr. Percival.

New York and New Jersey. — *New Red Sandstone.* — Messrs. Mather and Rogers.

Pennsylvania and Virginia. — *New Red Sandstone.* — Messrs. Rogers.

Lake Superior. — *Old Red Sandstone.* — Mr. Bayfield. *Potsdam sandstone.* — Messrs. Houghton, Logan, Foster, Hall and Owen. *New Red Sandstone.* — Messrs. Jackson and Marcou.

Rocky Mountains (near Devil's Gate). — *Conglomerate and sandstone?* — Mr. Stansbury.

Upper California and Oregon. — *Early sandstone and conglomerate?* — Mr. Dana.

The New Red Sandstone is generally composed of two parts; the lower is chiefly formed of thick beds of conglomerate, lying horizontally on metamorphic or eruptive rocks, or even on dislocated beds, with discordant stratification, of the Carboniferous, Devonian and even Silurian groups. This conglomerate has, as yet, presented no traces of fossils; and it is probable, if any should be discovered hereafter, they will be few in number, and in a bad state of preservation. These strata are especially remarkable for their points of junction with the copper trap, which dislocated them immediately at the end of the triassic period, and which, in some places, has penetrated and enclosed them in a net-work of dykes of native copper.

The upper part of the New Red is composed of beds of red sandstone with intercalation of thin beds of clay of the same color; the whole sometimes presenting gray and greenish tints, instead of red, which is, however, the prevailing color. The stratification of these beds is much more regular than that of the conglomerate. Numerous ripple-marks are often found at the surface of the strata, which seems to indicate that this deposit took place under the influence of very high tides, an opinion which is further confirmed by the numerous foot-prints of birds and drops of rain, so well described, by President Hitchcock, as found in the valley of the Connecticut. The Bay of Fundy offers remarkable examples of these phenomena, which are occurring at this time on all sandy and muddy beaches subjected to very strong tides; they have been described in detail by Messrs. Webster, of Kentsville, and Lyell, of London.

This part of the American Trias presents in certain places, as Connecticut and New Jersey, impressions of plants and fishes, which, although not very common in these localities, enable us to affirm, independently of the stratigraphic position of the formation itself, that this deposit belongs to the New Red Sandstone of England; or, more exactly, according to Mr. Agassiz, to the Keuper, or variegated marls, that form the upper part of the Trias of Europe.

Veins of fibrous gypsum, gypsum in masses, and springs of salt water, are often found in this formation, in New Brunswick, Nova Scotia, Prince Edward's Island, Magdalen Islands and Massachusetts. It is well known that it is the same formation in Europe that contains those immense beds of rock-salt and of gypsum that are worked on so large a scale in Lorraine, in Franche-Comté, in Switzerland, Würtemberg, Baden and Saltz-Kammergut.

The geographical distribution of the New Red Sandstone is limited to small basins, whose general direction is parallel to the dislocation of the Alleghanies, from the Magdalen Islands, in the Gulf of St. Lawrence, to North Carolina. In the Bay of Chaleurs, the Magdalen Islands, the Bay of Fundy, in the palisades of the Hudson River, at Lake Superior, and in the valley of the Connecticut, it often presents long lines of nearly horizontal beds, capped by masses of trap similar to the masses of basalt of Auvergne and Ireland, and, like them, divided transversely, and presenting the columnar structure so celebrated in the Giant's Causeway.

The whole of the southern, as well as many points on the northern coast of Lake Superior, present this formation; Point Keewenaw and Isle Royale are especially celebrated for the rich mines of native copper and silver contained in trap which crosses this formation.

According to Captain Stansbury's account, I am inclined to synchronize several beds of conglomerate, described by him in the environs of the Devil's Gate, Rocky Mountains, with the system of the New Red. I have, also, good reasons for thinking that the conglomerate and sandstone described by Mr. Dana, in his learned and beautiful volume of *The Geology of the Exploring Expedition of the United States*, and found by him on the Shaste River, and the boundary beween Oregon and California, belongs to this same formation of the American New Red Sandstone. It is with some hesitation that I advance these two last suppositions; and it is quite possible that later geological observers of those regions may change the age of this conglomerate of the Rocky Mountains and the Pacific.

The most common fossils of the New Red are:

Eurynotus tenuiceps, Agass. (Pl. VI. fig. 1, 1 *a.*)—This fish, of the order of the *Ganoids* of Mr. Agassiz, is found in abundance at Sunderland, Massachusetts. The genus to which it belongs is especially characteristic of the Keuper in Europe.

Ornithoidichnites, Hitchk. *Brontozoum Sillimanium*, H. (Pl. VI. fig. 3.)—The foot-prints of birds are found in great number in the valley of the Connecticut River. Their form leaves no doubt that they were made at the epoch of the deposit of the New Red Sandstone, by the birds that lived in that geological period.

Impression of rain-drops, Redfield. (Pl. VI. fig. 2.)—This preservation of the impression of rain-drops on muddy beaches at low tide is certainly one of the most curious facts which geology presents to our notice, and it shows the diversity of the medals of the creation, where everything is preserved, from the Mastodon to the insect and the drop of rain. These rain-prints are found at Pompton, New Jersey, in the Connecticut valley, and at Lake Superior.

LIAS OR JURASSIC GROUP.

For a long time the Jurassic system was not considered to exist in America; and it has only been positively recognized within a few years past. It is true that it occupies a very restricted space, when compared with the other systems, in those portions of the country that have been geologically explored. It is probable that further knowledge of the geology of the countries yet unknown will increase the extent, and, in so doing, the importance of the Jurassic rock, in the New World; for, until now, it has only been recognized in Chili by M. Domeiko, in Virginia by Mr. W. B. Rogers, and in the Rocky Mountains, Raton Mountain, and Muddy River, by Messrs. Frémont and Abert.

The most remarkably characteristic feature of the Jurassic in North America is a great development of beds of soft coal, entirely similar to the soft coal of the carboniferous system; but the fossil plants which are found in the clay and sandstone intercalated with the coal belong to genera and species entirely different from those which compose the carboniferous flora of the palæozoic epoch. They have the closest analogy, however, to the fossil plants of Europe; and some are even identical with those which compose the Triassic and Jurassic flora of England, Germany and France. Moreover, M. Agassiz has found fossil fishes, belonging to genera exclusively liassic in Europe, which, added to the accidents of dislocation evidently posterior to those which have put an end to the period of the American carboniferous group, evidently indicates, for the relative age of these coal deposits, the secondary epoch,— more especially the Lias, which, as is well known, forms the first division of the Jurassic group of Europe.

The Liasic Formation of North America has been studied carefully and in detail only in the State of Virginia, where it forms a band from ten to twelve miles in width, which extends only from the environs of Richmond to the neighborhood of Alexandria and Washington,— that is, about fifty miles long. Enclosed in a sort of long granitic furrow, very deep and very narrow, the rocks which compose it are a coarse-grained sandstone, formed from the decomposition of the surrounding granite, a species of micaceous schist, often very clayey and passing into black slate; lastly, beds of coal placed chiefly at the lower part of the formation, forty or fifty feet in thickness.

According to the collection of fossil plants made by the officers of the United States army, the beds of coal which are found at Raton Mountain, on the route from Missouri to Santa Fé, and at Muddy River, on the route to Oregon, have been recognized as also belonging to the Jurassic epoch.

The names and figures of the four most common and most characteristic plants of the Virginian lias are given, of which two are also found in Europe:

Equisetum columnare, Brongn. (Pl. VII. fig. 2.) — This species, which is the most common in the coal basin of Chesterfield County, in Virginia, is also very frequent in Europe, where it is found at Whitley, England, Brora, Scotland, in the Jura, and in Wurtemburg.

Tœniopteris magnifolia, Rogers. (Pl. VII. fig. 1.) — This magnificent leaf, which attains considerable size, is found chiefly at Gowrie Mine.

Zamites vittarioides, Brongn. (Pl. VI. fig. 5.) — This plant, also known by the name of *Z. obtusifolius*, is still more abundant than the preceding, with which it is almost always associated.

Pecopteris Whitbiensis, Brongn. (Pl. VI. fig. 4.) — This species, though rare in Virginia, is nevertheless found at Gowrie Mine and at Creek Mine; but much less abundantly than at Whitley and Scarborough, in Yorkshire, England.

CRETACEOUS GROUP.

If the Trias and the Jurassic groups are very little developed in America, the chalk formation, on the contrary, covers a considerable extent of country, and occupies a place in the series of rocks at least as important as the Silurian system. Discovered first in New Jersey by Vanuxem, the cretaceous strata were soon recognized in Delaware, Virginia, the Carolinas, Georgia, Alabama, Louisiana, Mississippi, and Tennessee. Mr. Featherstonhaugh noticed them in Arkansas and on the Washita River; and, lastly, Nicolet, in his exploration of the Missouri, recognized cretaceous fossils on the shores of this river, from Council Bluffs to Fort Pierre Chouteau.

Notwithstanding the great number of points where the cretaceous group has been verified, no detailed description of the different divisions that compose this system in the United States, in harmony with present geological knowledge, has yet been given; and hardly anything has been added to the labors of Vanuxem and Morton, which were published in 1828 and 1834.

In New Jersey, which is the State where the cretaceous strata have been most carefully studied, two very distinct divisions are recognized. The lower, composed of greensand and green marl, is very analogous to the *upper greensand* and *marly chalk* of England, or *la craie tuffeau* of France, and contains fossils proper to itself, which are not met with in the upper division. This presence of Greensand, entirely similar, in a mineralogical point of view, to the Greensand of England and France (*Perte du Rhône*), is another new and curious proof of the constancy of the mineralogical character of the formations for the same geological epoch, and shows that, although this means cannot be used with certainty to identify one formation with another, it ought not to be neglected in the comparisons which are established, and that it is an aid which geologists ought to employ when they seek to establish the synchronism of the formations of two different countries.

The upper part of the cretaceous of New Jersey is chiefly calcareous rock, composed in great part of a soft, straw-colored limestone, easily disintegrated when it is exposed to the air, and having an entirely different lithologic aspect from any of the hard and compact limestones of the American palæozoic formations. This upper part, which corresponds, without doubt, to the *white chalk* of Europe, contains a very interesting fauna, and one which presents an entire contrast to those of the anterior systems we have passed through. The Molluscs, the Radiata and the Polyps, instead of presenting the forms of innumerable Brachiopods, Crinoids and enormous polyps, so characteristic and so remarkable, of the palæozoic epoch, are, on the contrary, represented by *Belemnites*, *Ammonites* and *Echinoderms.* The Trilobites have long since disappeared; and, instead of finding, as at Trenton Falls and Cincinnati, long *Orthoceras* and charming *Trilobites*, you find, at Bordentown and Timber Creek, *Baculites* with dentated partitions, and not simple, like the *Orthoceras*, the teeth of *Squalidæ*, and, lastly, numerous vertebræ of Saurians belonging to the genera *Mosasaurus* and *Pliosaurus.*

The two lower divisions of the cretaceous group of Europe, which are the *Gault* and the *Neocomian* or *Lower Greensand*, have not yet been recognized in the United States; nevertheless, it is highly probable that they will be discovered in the Rocky Mountains or the Indian territories of the far West. For the Neocomian has already been verified in New Grenada, as well as in the Cordillera of Quito; and, as I have said, the cretaceous system has only been studied in detail in some isolated fragments, scattered from New Jersey to Georgia.

By a glance at the distribution of the cretaceous group upon our map, it will be seen that, with the exception of small fragments situated in the Atlantic States, from the bay of New York to Savannah, the great mass of cretaceous strata is situated in the basin of the Gulf of Mexico; and that, from Fort Mandan, on the upper Missouri, to Presidio del Rio Grande, Mexico, and to Macon, Georgia, there is a long, uninterrupted band of rocks belonging to this formation. The eastern and southern limits may be regarded as certain; but it is not the case with the northern and western, where, except in very few localities, the rocks are not known with which the cretaceous group is in contact; and it is probable, owing to the great difficulty of exploring these waste regions at the foot of the Rocky Mountains, that it will be a long time before the northern and western limits of the American cretaceous group can be determined.

At the west of Santa Fé, an officer of the United States army has recognized this group of strata, very well developed; from which we may anticipate that it occupies a remarkable place in the dislocations of the Rocky Mountains, and that it will one day serve to fix the relative age of some of the chains which compose this immense mountainous mass which separates the Missouri from the Colorado and the Columbia.

The following fossils are the most characteristic of the cretaceous system of the United States:

Belemnitella mucronata, Blainv. (Pl. VII. fig. 3.) — This species of Cephalopod Mollusk is one of the most widely spread, both in Europe and America. It is found in New Jersey, Delaware, Georgia, Arkansas, Tennessee, Texas; in England, France (Meudon, near Paris), Holland (Mäestricht), in Sweden, Prussia, Russia, Poland, Georgia (Asia), Volhynia, etc.

Ammonites nebrascensis, Owen. (Pl. VII. fig. 4.) — This Cephalopod is found in the Fox Hills of Nebraska, a little to the north of Fort Pierre Chouteau. Almost all the cretaceous fossils which are found in the regions of the upper Missouri have preserved their primitive color.

Baculites ovatus, Say. (Pl. VII. fig. 5.) — This Cephalopod is frequently met with in New Jersey, Delaware, Alabama and Texas.

Inoceramus Sagensis, Owen. (Pl. VII. fig. 6.) This Acephalous Mollusk, which bears much resemblance to the *Pecten*, belongs to a genus that is considered, with reason, as characteristic of the cretaceous system. This species has been found at Sage Creek (Bad lands), in Nebraska.

Terebratula Harlani, Mort. (Pl. VII. fig. 7.) This species of Brachiopod varies much in its form. It is especially characteristic in New Jersey of the lower part or greensand of the American cretaceous system.

Hemiaster parastatus, Mort. (Pl. VII. fig. 8.) This Spatangoid is abundant at Timber Creek, New Jersey, where it characterizes the straw-colored limestone which composes the upper part of the formation.

CHAPTER III.

TERTIARY, QUATERNARY AND MODERN ROCKS.

THE Tertiary Rocks follow nearly the same line of geographical distribution with the Cretaceous, but are much more developed, especially along the Atlantic coast, where they often entirely cover the beds of the Cretaceous group, and where they come to rest directly upon the eruptive and metamorphic rocks.

Their study, like that of the cretaceous, is far from presenting as satisfactory results as that of the palæozoic rocks of this continent. Often confounded by observers with the cretaceous, quaternary, and even modern system, it is not possible to give a very exact general description of them, or to limit them with certainty upon the map; and I have taken the same color for the tertiary and quaternary rocks, because it is not possible to distinguish them accurately at a large number of points.

TERTIARY. — The only researches of any accuracy and detail upon this formation are those of Messrs. Conrad, Lea and Lyell, who have determined the relative age of several groups of these rocks, and in this way have established certain data for later students of the rocks of this system.

The tertiary rocks have generally been divided, in nearly all countries where they have been recognized, into three groups,— the lower or Eocene, the middle or Miocene, and the upper or Pliocene. Each of these groups has been again divided into several sub-groups in many regions of western Europe, where the tertiary system occupies a considerable place in the geographical extension of sedimentary rocks, and where it has been studied with great care and minuteness, which is not the case in America. For this reason, we shall only seek to establish the three principal formations.

The Eocene strata, or inferior tertiary, are found along the eastern border of the eruptive and metamorphic rocks of the Alleghanies, beginning at Delaware, forming a band which enlarges as you go southward, and whose distinctive character becomes very marked, especially in Alabama, Mississippi and Louisiana. In Maryland and Virginia the rocks which constitute the Eocene are principally green-sand and marl, having the greatest mineralogical resemblance to the cretaceous rocks of New Jersey; which is easily explained by a partial destruction, and a residuum of the cretaceous system at the moment when the formation of the Eocene began.

In the Carolinas, Georgia and Alabama, this mineralogical character of the Eocene changes, and is replaced by a white limestone, often very compact, with red and white clays; and sometimes, especially in Texas, by ferruginous sand. The fossils found in these strata present a very different aspect, in general, from those which form the cretacean fauna; instead of cephalopod mollusks, such as the genera Ammonites, Belemnites, Baculites, &c., there are a large number of acephalous and gasteropod mollusks belonging to the genera Ostrea, Lucina, Natica, Fusus, &c.

Many of the Eocene fossils are identical with, or very much resemble, the species of Europe, and give an age to these American strata which is probably the same as that of the calcaire grossier of Paris or of the nummulitic rock on the borders of the Mediterranean. The localities of Fort Washington, Maryland; Richmond, Virginia; Wilmington, North Carolina; Santee River, South Carolina; Jacksonboro', Georgia; Claiborne, Alabama; Vicksburg, Mississippi, are celebrated as types of the American Eocene. Claiborne especially is the richest locality in fossils, and has been the most studied.

Of late the Eocene strata have been recognized in the north-western regions, with a very marked character of fresh-water deposit, wholly analogous to what is found in the basin of Paris. Messrs. Stansbury, Culbertson, Evans and Richardson, have collected fossils, mollusca as well as plants, and especially vertebrata, which leave no doubt of the presence of a large Eocene basin, extending from the borders of the Platte River and Republican Fork, through the Bad Lands, the Upper Missouri and the Sakatchewan River, to the mouth of the Mackenzie River, in the Arctic Ocean. The knowledge of this immense basin is still too imperfect to allow us to affirm that it is wholly composed of

5

a fresh-water formation, but the organic remains which have as yet been discovered by Messrs. Richardson, Prout and Leidy, are of a lacustrine, fluviatile, or estuary character. Mr. Römer has determined the junction of the Eocene strata with the cretaceous rocks in Texas. It follows a line, which, starting from the Rio Grande del Norte, between Matamoras and Laredo, ascends nearly to Presidio del Rio Grande, passes to San Antonio de Bexar, Austin, Fort Towson, and reaches nearly to Little Rock, in Arkansas.

The middle and upper formations of the Tertiary system have been recognized and studied still less, and on fewer points, than the lower formation. The Pliocene, especially, has been recognized only in two or three localities in Virginia and New Jersey; and it has probably been confounded in many places with the rocks of the Quaternary system of the Miocene. As yet, the existence of Miocene and Pliocene has not been certainly indicated in the States of Mississippi, Alabama or Georgia. It is only from South Carolina to the sea-coast of Maine that beds of sand and clay have been described, containing fossils identical with, or very similar to, those found in the molasse of Switzerland, the faluns of Touraine, France, and in the crag of Suffolk, England. The principal localities where the Miocene and Pliocene strata have been recognized are Wilmington, North Carolina; City Point and Coggin's Point, Virginia; Williamsburg, Maryland; Cumberland County, New Jersey; West Point, New York; Nantucket and Martha's Vineyard Islands, Massachusetts; the Connecticut valley, and Portland and Augusta, Maine. It is very probable that the States which surround the Gulf of Mexico contain tertiary rocks belonging to these middle and upper divisions; and further, that these rocks ascend the Mississippi as far as Tennessee and Arkansas.

In comparing the following tertiary fossils with those of the other geological periods contained in this volume, the great difference of their forms will be seen, and at the same time it will appear how much these tertiary fossils resemble the shells actually living on the shores of the Atlantic and the Gulf of Mexico.

Lucina rotunda, Lea. (Pl. VIII. fig. 1.) — This Acephalous Mollusk is found in great abundance at Claiborne, Alabama; and on the Santee river, South Carolina

Venericardia Sillimani, Lea. (Pl. VIII. fig. 2.) —This Acephalous Mollusk is also very common at Claiborne, at Wilmington, and at Shell Bluff, Georgia.

Ostrea semilunata, Lea. (Pl. VIII. fig. 3.) — The genus Ostrea is one of the most universal and characteristic of the American tertiary epoch. This species is found especially at Claiborne and Santee River.

Natica striata, Lea. (Pl. VIII. fig. 4.) — This little Gasteropod Mollusk is one of the most characteristic fossils of Claiborne and Wilmington.

Fusus Fittoni, Lea. (Pl. VIII. fig. 5.) — This gasteropod, which resembles closely a species found at Grignon, near Paris, is found at Claiborne and in Virginia.

Voluta Vanuxemi, Lea. (Pl. VIII. fig. 6.) — A species of Gasteropod, very common at Claiborne, where, with the five preceding species, it characterizes the Eocene strata.

Charcharodon angustidens, Agass. (Pl. VIII. fig. 8.) — This tooth, which belongs to a fish of the family of the Squalidæ, is one of the most characteristic fossils of the American and European Eocene, and is quite common.

Oreodon Culbertsonii, Leidy. (Pl. VIII. fig. 7.) — This fossil Mammal, of the order of the Pachydermata, was truly a ruminating hog. A great number of specimens have been found in the Eocene rocks of Nebraska, in the Bad Lands, between Cheyenne and White river, to the north-west of Fort Pierre Chouteau. This part of the Nebraska Territory appears to be richer in the remains of Mammalia and Chelonia of the Eocene period than the deposits of the same age, in the basin of Paris. All the fossil mammalia which have been found there belong to the order of the Pachyderms, with the exception of one carnivorous animal of the genus *Machairodus*. The most common genera are *Palæotherium*, *Rhinoceros*, *Oreodon* and *Archæotherium*; these two last are entirely new genera in Paleontology, and appear to belong exclusively to the Eocene fauna of North America.

QUATERNARY. — The quaternary system, or diluvium, includes all deposits, whether regularly stratified or not, — marine, fluviatile, lacustrine, terrestrial, solid or incoherent, — which were formed between the close of the upper tertiary period (Norfolk and Suffolk upper crag) and the commencement of the modern rocks, or the present epoch.

These rocks, which correspond to the groups *Newer Pliocene or Pleistocene* of the tertiary, and *Post-Pliocene of the Post-tertiary* of Mr. Lyell, are even now very difficult to characterize, notwithstanding the numerous works lately published on the geological period. Confounded at one time with the tertiary and at another with the recent deposits, the only writer who has classed and limited the quaternary rocks with an approach to regularity is Mr. D'Archiac, in his important work, entitled "*Histoire des Progrès de la Géologie.*"

"The indistinct character," says D'Archiac, "of the deposits which the quaternary epoch has left us, thinly spread over large surfaces, and the want of regularity, symmetry and continuity, in their general disposition, have rendered incomplete and uncertain all attempts to compare them, and establish relations between them."

Notwithstanding several examples have been cited of human remains, or of the relics of man's industry, as found in the caverns and deposits of diluvium, together with bones of species of vertebrata entirely lost, very few geologists have been convinced that they were contemporary; and it may be said that the precise epoch of man's arrival on the earth is still a profound mystery, into which, perhaps, we shall never be permitted to penetrate.

The formations which belong to the quaternary are much more complicated than was at first believed; and the theorizers who explained them all, some fifteen years since, by simple and easy means, have seen their theories vanish and disappear behind a mass of newly-observed facts, which, unhappily for the theorists, resist these explantions *à priori*, always so convenient, but also always false, when they are applied to the whole globe, or even to one of the two hemispheres.

In America these formations are composed of raised beaches of marine and lacustrine deposits, and, lastly, of drift and boulders, or erratic formation. On the line of coast extending from the mouth of the Hudson River, at New York, to the mouth of the Rio Grande del Norte, in the Gulf of Mexico, are found ancient beaches, often fifty or even a hundred feet above the present level of the ocean, which contain, in their strata of sand and clay, shells in a fossil state, exactly the same as those that are now living in the sea. Ascending the rivers of this region, on almost all the cliffs, and especially on the borders of the immense basin of the Mississippi, sandy and clayey deposits, containing enormous quantities of various fresh-water shells, such as the *Unio*, *Anodon*, *Helix*, *Pupa ;* of species that now live on the borders or in the bed of these rivers. The cliffs containing these shells are often raised more than a hundred feet above the rivers which they surmount; on the heights overlooking Pittsburg, between the Monongahela and Alleghany Rivers, there are beds of clay containing great numbers of the Unio, with the same variety of species that exist in the actual Ohio, and which are one hundred and twenty feet above the level of the river. There are often found in these strata numerous bones of the mastodon, elephant, tapir, wolf, megatherium, mylodon, &c., in a good state of preservation; all those mammalia and gigantic chelonians which have preceded the actual fauna. The most celebrated localities for these bones are San Felipe, Texas; Natchez, Mississippi; West Feliciana, Tennessee; Bloomfield and Big Bone Lick, Kentucky; Cincinnati and Zanesville, Ohio.

Many of these enormous skeletons of mastodons and elephants have been discovered entire; and, notwithstanding the difficulty of preserving them,—for they often crumble into dust on coming into contact with the atmosphere,—several are found in public and private museums. The Mastodon giganteus (Cuvier), found at Newburg, New York, which belongs to Dr. J. C. Warren, is celebrated for its fine preservation, and by the engravings which have been made of it by its proprietor.

On the heights which border the St. Lawrence from its mouth to its source, that is, to Seven Beaver Lake, the source of the St. Louis River, the first name of the St. Lawrence, varying from fifty to one hundred feet, are found deposits of sand and clay, often forming terraces above the river, which contain shells identical with those that live now in the waters of the River and the Gulf of St. Lawrence. Lake Champlain and the River Richelieu, also, present the same phenomena. The difference between these deposits of sand and clay and those we have described as further south is, that they overlay a formation often considerable, composed of drift, boulders and scratched and polished rocks, which is special to the polar regions, or to the high mountain chains of the temperate and equatorial zones.

This formation of drift and boulders is chiefly marked by sand and clay, enclosing blocks which vary from the size of a pebble to that of an enormous rock, and which have been transported various distances from the point where they were formed. A constant and particular phenomenon of this formation is the marks of the passage of these blocks and gravel upon all the rocks that form the country where they are found. These marks consist of scratches, often very fine, the majority of which follow a certain direction; although in America a surface is seldom found with all the scratches parallel, which sometimes occurs, within narrow limits, in the Alps. The scratches occasionally cross one another at all angles, from zero to ninety degrees, though they have still a general direction, which the crossing does not change; and they are imprinted with the same regularity on all sorts of rocks, even the conglomerate, which are composed of fragments of various hardness; which shows that the force that produced them must have been uniform and powerful.

In the researches lately made into the cause or causes that produce these scratches, it has been determined that ice only has this power, and it has been concluded that the drift and erratic boulders owed their origin to the glaciers or to floating ice. In North America, all the immense erratic formation, extending from the pole to the forty-first degree of latitude, has been attributed to glaciers by some writers; but I think the country is too flat, and the constant intersection of the scratches cannot be explained by the motion of glaciers. Consequently it has been concluded, and I think justly, that the greater part of the American drift and boulders are due to icebergs and ice cakes, still so common now in Lake Superior, on the coast of Labrador, the banks of Newfoundland, in Hudson's Bay, &c. If there are real glacier scratches, they will be found in the Rocky Mountains, and perhaps also in the White Mountains.

On glancing at our map, it will be seen that the quaternary system is indicated at very few points. Long Island, Cape Cod and the Plateau du Coteau du Missouri, are colored as belonging to this system, which I have made the same tint as the tertiary. Very few points in the United States are free from more or less considerable traces of the quaternary system; thus I have not indicated it where the age of the rocks on which it rests is known, whatever might be its importance in extent and thickness. To have a just idea of its distribution and its true relations, a special map would be needed, presenting only the tertiary, quaternary and modern systems, geologically colored. Unhappily, many of the elements are still wanting to such a a study.

MODERN ROCKS.

The modern rocks are divided into two large classes. The first comprehends all the products of the causes which tend to modify the surface of the earth by acting directly upon it; and the other embraces those phenomena which have their origin below the earth's surface.

The products or results belonging to the first class are, according to their origin, atmospheric and terrestrial, such as the alteration of rocks by atmospheric action, deposits produced by falling dust or ashes, vegetable soil, landslips and debacles; also lacustrine, fluviatile and marine, such as icebergs, glaciers, the alluvium of rivers and lakes, deltas, sea-margins, dunes, coral islands and reefs, &c.

Those of the second class are aqueous, such as mineral and thermal springs, tuffs and travertine; or volcanic, as volcanoes, earthquakes, contemporary elevations and depressions.

In a word, the modern rocks are the result or products of all the phenomena now in action in or upon the planet named Earth. The study of these rocks is of the greatest importance, as it first acquaints us with what is passing around us, and then enables us to compare and unite, by the laws of analogy, the present phenomena to those which have preceded them.

Although we regard the study of the present causes and results the most certain basis that the geologist can command, to explain in a rational manner what has passed upon our globe, yet we are far from thinking that the organic and inorganic phenomena now in action, however much prolonged their action be supposed, can explain all those which took place during the geological periods, and all the effects which are produced by them.

The school of actualists proceeds by negation. It denies all the researches, paleontological, mineralogical and geognostic, which have been made upon the epochs anterior to the tertiary system, and substitutes in their place doubts and prophecies. Thus, the progressive perfection in the order of the vertebrata, from the Silurian to the present epoch, is for the actualists a myth; the complete annihilation of one fauna, and its replacement by another wholly different, is impossible, and contrary to the existence of Noah's ark. But, in recompense, the actualist predicts that a fossil man will be discovered in the Potsdam sandstone; that the sea-serpent, so celebrated in the newspapers, is a lineal descendant of the Ichthyosaurus or Plesiosaurus of the lias of Lyme-regis; and, finally, that you may some day see granite, dykes of gold and silver, and beds of cannel coal, formed before your eyes.

Happily, geology continues its progress, notwithstanding the barriers that theories would impose. In the natural sciences, it may be said, without hesitation, after the efforts of Linnæus, Werner, Hutton and Geoffroy St. Hilaire, that every theory is false; that the earth does not bend to the laws of mathematics, like the heavenly bodies; and that man is incapable of explaining and classifying, under one law, the work of Him who has presided over the creation and development of our planet.

Geology, the résumé of the natural sciences, escapes still more easily than zoology, botany and physical geography, the limits of a system of classification. To submit not only geology, but one only of the numerous parts of which this science is composed, to a general theory, whose law is applied to the whole earth, is as impossible as to resolve the problem of squaring the circle, or of perpetual motion.

An order of geological facts in a certain geographical region may be explained in a more or less satisfactory manner; and that ought to be, and is truly, the aim of the researches actually made by the great majority of geologists. But, to deduce a law from a small number of observations, made on a twentieth or a thirtieth part of the surface of the globe, to which the whole shall be submitted, is an impossibility on which all the theorists, past, present and future, will make shipwreck. Notwithstanding the flexibility of the exterior part of the earth, no one is permitted to mould it according to the ideas of his brain. God only lays his hand upon it, and modifies or forms it according to his infinite and impenetrable designs.

But, to return to the description of the modern rocks. In North America the modern formations are produced on a large scale; their origin, as given above, will suffice to show their importance. I shall not describe in detail each rock actually in process of formation: only observing that there are no volcanoes in activity in the United States or in the British Provinces; that the phenomena of icebergs and ice-cakes take place annually on the great lakes, the St. Lawrence, the banks of Newfoundland, &c., and that the sea-margins extend chiefly from Long Island to the mouth of the Rio Grande del Norte. Several mud deposits present very curious phenomena, especially the red mud constantly deposited by the tide in the Bay of Fundy, in the Minas Basin. During the drought of summer, this mud is hardened before the return of the tide, and pieces of dried mud can be cut, showing on the edge the thickness of mud brought by each tide,—a thickness varying from four to eight lines; and if these little beds are separated, they often show the prints of birds' feet, of turtles, and the tracks of mollusks, even the drops of rain; thus explaining the similar marks found in geological periods anterior to ours.

The Ohio, Missouri, Red River and the Mississippi, bring down large quantities of organic and inorganic bodies, which deposit themselves on the coast of the Gulf of Mexico, and form an immense delta at the mouth of the Mississippi. Each year this delta encroaches on the ocean, and the bar, off the mouth of the river, continually changes place. I have indicated on the map this part of the shores of the Mississippi, which is inundated nearly every year, and offers deposits of modern rocks.

One of the most remarkable of the deposits of modern rocks is doubtless the point which forms Florida, which is entirely owing to coral banks, gradually but constantly increasing, and added to the coralline shore of this peninsula. This discovery of the recent formation of Florida, and the description of the mode of growth of the North American coral reefs, is due to Mr. Agassiz, who has given an excellent description of them in one of the publications for the United States Coast Survey, in the year 1852. I shall only give here the two polyps which play the principal part in the increase of the coral reefs. They are the Meandrina labyrinthica, Lam. (Pl. VIII. fig. 9); Astrea mammillata, Lam. (Pl. VIII. fig. 10).

CHAPTER IV.

GEOLOGICAL MAP AND SECTIONS.

THE first care of an observer, who makes a geological study of any country, is to draw sections which show the order of superposition and the arrangement of the strata; then, after having grouped these strata, he endeavors to limit them, and to trace, on a map, the boundaries of these different mineral masses.

The strata, or, more generally, the mineral masses, are not grouped by an absolute and mathematical law, but their arrangement depends upon the object proposed, and may vary even, according to the purpose of the observation which the geologist has in view. Thus, for example, in this work, whose object is chiefly to synchronize the sedimentary rocks of America with those of Europe, I have employed, to form my groupings, the three following means: first, the analogy, and even the identity, of fossils; then, the mineralogical composition of the rocks with regard to their general resemblance, and, lastly, the direction of the systems of mountains.

The last two means are of much less value than the first, but should not be entirely neglected, especially when the first is doubtful or wanting. But, if a topographical map, on a large scale, of one of the States of the Union, were to be studied in detail and colored geologically, the synchronism of the mineral masses contained in this State with those of Europe being only accessory to the principal object, I should make the grouping of the strata chiefly according to the state of aggregation of the rocks; thus, as they are sandstones, clays, limestones, or schists, &c., the topography of the country will correspond to them and result from them, taking into view, nevertheless, the manner in which these rocks have been affected by dislocations or denudations. Paleontology here is of secondary importance, and serves to show only that the faunæ and floræ buried in the strata are in harmony with the movements of the earth's crust which have given birth to the country. Lastly, the discordance of the stratification is also a characteristic of great value in such a study.

A geological section, when it embraces only a small and smooth extent of country, with the same scale for the heights as for the lengths, is of the greatest utility to give an exact idea of the superposition of the strata, the direction of dykes, and the position of the rocks with relation to each other. But, when the geological section crosses a chain of mountains, and includes a large extent of country, it is only a miserable caricature of the truth. For, in this case, the heights are considerably exaggerated in proportion to the lengths, in order to mark the strata which are exposed to view in these mountains, whose relative position it is requisite to show. Further; the curving of the earth is a new source of error, to which it is not usual to pay attention, but which does not fail to lead astray when the section crosses a large basin. Thus, when a geologist constructs or studies a large geological section, his attention should be constantly fixed on the proportions which exist in the section, as well as on the relation of these proportions to the dimensions of the globe.

I have added to the geological map of North America two large sections, which cross it in different directions, almost at right angles to each other. One of these sections goes in a straight line, from Yorktown, Virginia, to Fort Laramie, Rocky Mountains, crossing the Alleghanies perpendicularly; the other begins at Lake St. John, Hudson's Bay Territory, and follows a broken line, which passes by Lake Simcoe, Cincinnati, Nashville, and Mobile.

These sections are far from representing the reality, for the heights are exaggerated in proportion to the lengths, and I have not taken into account the curve of the earth. But, notwithstanding the want of exactness in their execution, I thought it best to give them, because they express that the chains of mountains have profound roots in the depths of the earth, and that the disposition of the plains is in harmony with that of the mountains. Further; these sections are a real assistance, enabling the student to read with more facility what the geological map represents. Nevertheless, I repeat it, that the reader may understand the graphic exactness of the sections, they are, in truth, caricatures; though I have endeavored to keep as near the reality as possible, and I think they express well enough what they are designed to express,—that is, the superposition of the rocks and their relative distribution. I have said these sections *express*, and do not *represent*; for they are truly only an expression, and not a representation or portrait of things as they exist in reality.

A topographical map, geologically colored, and accompanied by an explanation of the colors, arranged in the chronological order of the rocks, gives a much better idea of the exterior configuration of the globe and of the relation of the different mineral masses, than a geological section can do, even when the proportion of the heights to the distances is observed, and the curve of the earth is considered. Topography is one of the most important points for a geological map; for, without it, it is materially impossible to describe and understand the orographical structure of a country. The geological structure of the Alps, the Jura, and Auvergne, have been exactly described and understood, because there are good topographical maps of these regions. Unhappily, America has as yet no good topographical maps; and, notwithstanding the attempts at orographic description which have been made upon the Alleghanies, the degree of exactness of these theories cannot be determined, nor to what part of these mountains they may be applied, except from the sections made by geologists who have been engaged in this study.

In the geological map here given, there are no topographical indications, for the very simple reason that, being unable to make them exact from the want of topographical maps, I preferred to omit them entirely. But I have written near the places occupied by different chains of mountains the names of these chains, in order to give an idea of their relative position, and, also, that I may class these mountains according to their relative age; that is, following the chronological order of their appearance.

ERUPTIVE AND METAMORPHIC ROCKS.

DIRECTIONS OF MOUNTAIN RANGES.

Hitherto we have described the stratified rocks whose successive beds cover the larger part of the surface of the United States and of the Provinces. There is a second category of unstratified rocks, whose masses, coming from an unknown depth in the interior of the globe, rise, like irregular columns, and reach the surface, generally forming the summits of the highest mountains.

These rocks, of igneous origin, have been inserted in the fractures of the stratified rocks by eruptions from the interior of the earth, and have given rise to dislocations, and often to alterations, of the beds of aqueous origin.

The causes of the formation and appearance of the eruptive rocks are still theoretical questions, and this is not the place to discuss them. I shall indicate only their geographical distribution in North America, endeavoring to class them according to the epoch at which they rose from the bosom of the earth.

The igneous rocks are divided into a large number of different species, if they are classed according to their chemical and mineral composition. We shall name only the principal ones, referring to the appendix for their description. They are granitic, felspathic, hornblendic, serpentinous and porphyritic, according to the predominance of one or several minerals in their composition. At their point of contact with the sedimentary rocks, it often happens that the latter are changed and metamorphosed from their first nature into mica slate, gneiss, crystalline limestone, &c. Notwithstanding the want of agreement among geologists as to the rocks which should be regarded as metamorphic,—some even going so far as to place granite among them,—it is very important on a geologic map to distinguish the metamorphic from the eruptive rocks, in order to take advantage of their relative positions, for orographic description, or for the age of the rocks that form a country. We have in vain attempted this separation in our map of America, the want of success being caused by the insufficiency of observations on many points, and by the inextricable confusion of the descriptions and classifications of them which have been made. For these reasons, the metamorphic and eruptive rocks have only one color, the copper-trap only being distinct, whose mineralogical character, position and exterior form, make such a contrast with the other igneous rocks, that it is easy to indicate this division of the igneous rocks by a special color.

It will be seen at once, by following on the map the parts of the country wholly formed of eruptive and metamorphic rocks, what an important part they play, first, as to extent, and then chiefly it may be remarked that they generally form the mountainous parts of the country, and almost always make the contours of the hydrographic basins. These last considerations, one of which is the corollary of the other, have, for a long time, attracted the attention of geologists; further, the practical examination of mountains has taught miners, from time immemorial, that the sedimentary beds are elevated according to general direction, whose constancy often extends to immense distances, and that the mineral veins also present this constancy of direction, to which is added a special parallelism in the veins of the same nature. Such results have fixed the attention of observers, and very marked relations have been recognized as existing between the direction of the mountains, the rocks which form them, and those that surround them. These relations are far from being all known, or even perfectly limited; and, notwithstanding numerous researches on this subject, it may be said that this part of geology, which treats of the origin, age, and orography of mountains, is the most obscure, and that upon which the science has the fewest facts. The most distinguished geologists, from the time of Werner and De Saussure, have occupied themselves with these questions; but no satisfactory results have yet been reached. At first regarded as not a difficult question, they sought rather to explain the origin of these relations, than to study, measure, and define them with exactness. In this way the number of theories conceived and uttered on the subject is considerable; but, as observations have increased, the question has been recognized as a complicated problem. And, now, instead of constructing theories on these relations, practical geologists are satisfied to describe them, hazarding timidly, from time to time, a theoretical suggestion, very far from having the absolute and mathematical exactitude which was formerly given to them.

The most plausible and rational of the hypotheses on the origin of mountains appears to be that of the gradual cooling of the globe, and the consequent contraction of the internal mass. On the relative age of mountains,—that is, the epoch at which each appeared,—satisfactory results have been obtained for a part of Western Europe, thanks to the labors of Messrs. Luch, Sedgwick, Bouè and Elie de Beaumont. This part of the science, which may be designated under the name of Geodesic Geology, is still far behind the classification and study of the sedimentary and igneous rocks; and it may even be said that the methods to be followed, in order to arrive at the knowledge of these relative ages, are still very incomplete.

As to orography, which is, properly speaking, the geological topography of a chain of mountains, it is not yet known even for the mountains of the countries which have been the most explored, as England, France and Germany. Various theories have been proposed, based always on a small number of facts, not only to explain the orography of one chain, but of all the mountain chains of the globe. Unhappily for their authors, these ambitious orographic theories vanish still more rapidly than those which are made on the other parts of geology; and, even now, not only are there no general results obtained which may be applied to several chains, but there are satisfactory observations only on two chains; which are the Jura, by M. Jules Thurmann, and the Swiss Alps, by M. Bernard Studer; and even these orographic results of these two chains are far from explaining all the forms and accidents of dislocation found there. Thus it may be said, without hesitation, that to find a description of the structure of a chain of mountains, more or less in agreement with the reality, is the most difficult question for a geologist to resolve. In order to show still more clearly the difficulty of such a problem, it is sufficient to recall the endless discussions to which the geognostical orography of volcanoes has given rise, this phenomena of isolated mountains which takes place before our eyes, and which may be called simple, when compared with the phenomena developed in the Alps, the Caucasus, the Himmalaya, the Rocky Mountains, &c.

The constant object of the geologist's researches is to make himself master of the secrets of nature.— How to observe? What steps to take in these interrogations that man addresses to the witnesses of what has formerly taken place upon the globe? "That is the question!" Each sort of observation requires special means and methods, but certainly none demands more sagacity, experience, I might almost say genius, than to be able to know the internal structure of a chain of mountains. It is the most hidden secret that geologists have yet been able to discover.

The fossil remains are the "medals of creation," while the mountains are its *colossal statues;* with a collection of a certain number of fossils and minerals, one may write, and even stereotype, a chapter of the world's history; but this chapter always begins with an immense *capital letter*, which is a chain of mountains.

As was remarked above, very little is known on the relative age of the chains of mountains of the globe, and the methods pursued for this object are still very uncertain. M. Elie de Beaumont is, certainly, the geologist who has contributed the most to advance this part of the science; and he has recently published his researches on this subject in a remarkable book, entitled, *Notice sur les Systêmes de Montagnes*. I will indicate, in a few words, a part of the principles and the method followed by M. de Beaumont:

"The fissures made in the external crust of the globe have determined the elevation and upheaval of the beds of which this crust is composed, and the edges of these broken and upheaved beds have become the crests of those irregularities of the surface of the globe called mountain chains; from which it results that the expressions, average direction of a system of fractures, average direction of a system of upheaved beds, direction of a system of mountains, are *nearly* synonymous.

"The circumstance that in each country the inclined sedimentary beds, and the crests which these beds form, do not present indifferently all sorts of directions, but conform themselves to a limited number of general directions,— a circumstance of which all maps exactly drawn present striking examples,— has appeared to me to constitute in the study of mountains a fact of an importance analogous to that which the fact of the independence of the formations presents in the study of the successive sedimentary deposits. I have endeavored to connect these two facts together; and I think I have established their coincidence in a sufficiently large number of examples to warrant the conclusion of the independence of systems of mountains having a different direction.

"The indication of a general tendency to parallelism, presented by the foldings and fractures of the earth's crust produced at the same epoch, seems, at first, to need no commentary, especially when applied only to accidents observed in the soil of a country so limited in extent that the curve of the earth can have little effect. Nevertheless, as no limit can be put to the distance to which it might be possible to follow accidents constantly obeying the same law, it is necessary to analyze this first notion of a certain parallelism with sufficient exactness, to prevent the extent of space over which this parallelism may exist from making the definition erroneous.

"To do this, in the first place, we must remember that when a line is marked on the surface of the earth, with a cord, or by stakes, or in any other manner, the line thus determined is the shortest that can be drawn between the extreme points at which it stops; and that the slight flattening which the terrestrial spheroid presents being abstracted, such a line is always an arc of a great circle.

"Two great circles which intersect each other necessarily at two points diametrically opposite can never be parallel, in the ordinary sense of the word. But two arcs of great circles, sufficiently limited in extent for each to be represented by one of its tangents, may be considered as parallel, if two of their respective tangents are parallel to each other. It is thus that all the arcs of meridians which cross the equator are really parallel to each other at the points of intersection.

"In general, two arcs of great circles of limited extent, even without being very small, may be called parallel to each other if they are placed in such a manner that a third great circle cuts them both at right angles through their middle point. For the same reason, any number of arcs of great circles, each of little length, can be called parallel to one great circle of comparison, if each, in particular, conforms to the condition above stated with regard to a tangent of this auxiliary great circle. For this, it is necessary and sufficient that the different great circles which cut at right angles each of these little arcs in the middle should themselves meet at the two opposite extremities of the same diameter of the sphere. If this condition is fulfilled, and if, at the same time, all these little arcs of great circles are separated from the two points of intersection of their perpendiculars,— if they are concentrated in the neighborhood of the great circle which serves for the equator to these two poles,— they may be considered as forming on the surface of the sphere a system of lines parallel to each other.

"The fundamental problem presented by a like system of small arcs observed on the surface of the globe, where they are marked by the crests of mountains, or by the outcrop of beds, consists in determining the great circle of comparison, to one of the elements of which each of the small arcs observed is parallel.

"The small arcs determined by observation may be generally considered as being themselves infinitely small secants, or tangents, to so many small circles resulting from the intersection of the surface of the sphere with planes parallel to the great circle of comparison, forming the equator of the whole system. Each of these small circles is a parallel with respect to the equator of the system; it has the same poles, and these poles are the two points where all the great circles perpendicular to the small arcs, constituting the system of parallel traces determined by observation, intersect.

"The problem arising from such a system of parallel traces observed on the surface of the globe consists in determining these two poles, or, what amounts to the same thing, its equator,—that is, the great circle of comparison to which each of the small arcs observed may be considered as parallel. This determination would be easy, and might be made after two, or at least a few observations, if the condition of parallelism was rigorously satisfied. Since, however, this, in general, is but approximatively accomplished, the determination of the great circle of comparison can only follow after numerous observations, well combined with each other; and thus, while the observations are not very multiplied, or spread over a wide space, we can advance towards this determination only by successive approximations."

M. de Beaumont employs two methods,—the one graphic, the other trigonometric,—in order to determine the great circle of comparison for the direction of a system of mountains. These methods, much too complicated to be here described, have enabled the author already to determine a certain number of directions of mountain ranges which traverse western Europe. These different systems of mountains, when traced on a terrestrial globe, are seen to form a part of a vast system of parallel chains, which extends far beyond the countries whose geological structure is known. As the small parallel chains have been recognized, by degrees, to be generally contemporaneous in all the parts of each of these systems which are situated in the well-explored regions of Europe, M. de Beaumont concludes that this law should be universal; and, until direct observations have shown the contrary, he regards each of these vast systems as owing its origin to one and the same epoch of dislocation.

With this for the basis, the author prolongs the systems of mountains which he has recognized in western Europe into the other parts of the world; and, in this way, a certain number of the mountain systems of America enter into his classification. Without being equally sanguine with M. de Beaumont as to the generality of this law, not only for countries out of Europe, but for the whole of Europe itself, we recognize, nevertheless, that such a parallelism of small contemporary chains exists in all circumscribed countries; and, further, that this parallelism with the contemporaneity extends sometimes even to America; but, in this case, it is the exception, and not the rule. We think the question is far from having been resolved; and, above all, we cannot admit, without large reservation, so general a law, for so immense and complex a phenomenon. However this may be, we will try to class the dislocations that have affected the American strata included in our map, by means of this parallelism of arcs of a circle, and that independently of parallelism with the mountain chains of Europe, and then we will see if some of them coincide with the systems of mountains traced in Europe by M. de Beaumont.

But, first, I would remark that the direction of the great chains should always be examined and determined on a globe; for many directions that appear in a right line on a planisphere will be curved on a chart of Mercator's projection. It is also necessary to establish a distinction between chains not parallel having the same direction; that is to say, when each one takes its direction according to its own north, with relation to parallel chains bearing according to the card of direction for the points of the compass given upon the map.

The following classification is entirely temporary, and only a first, imperfect attempt in a kind of research which has hitherto almost escaped the attention of the geologists who have described America. For, with the exception of some excellent observations made by Messrs. Jackson and Hitchcock, on the general direction of the rocks in the New England States and in Nova Scotia, no general consideration has been given to this subject, nor any classification indicating the geological age of these directions.

1. *Lawrentine System.* — The granitic, syenitic and gneiss rocks, which make the foundation of the Lawrentine Mountains, are affected with numerous dislocations that have uplifted them in different ways. These dislocations are not all of the same epoch; nevertheless, there is one main direction which prevails much over the other directions,

and is almost from east to west, with an average deviation of nearly 5°, which gives for the direction E. 5° N. and W. 5° S.

These systems of dislocations are the most ancient I have examined. I regard them as anterior to the deposition of the first beds of Lower Silurian; that is to say, previous to the formation of the Potsdam sandstone.

Being the most ancient, it follows naturally that these primitive dislocations, which form in truth the mass of the Lawrentine Mountains, have been subjected to much alteration by the crossing of the directions of subsequent dislocations, which, added to the great difficulty of exploring the country where they are found, renders the study of them not easy. The localities where this older system of dislocation of the Lawrentines may be best observed are, the northern side of Lake Superior, between the factories of Michipicoten and the Pic; the northern coast of Lake Huron, between French river, Lake Nipissing and Fort La Cloche; the line which goes from Lake Simcoe a little to the north of Kingston. I consider as of the same system of dislocation the centre of the group of mountains, of eruptive and metamorphic origin, that forms the high land of Wisconsin and of Michigan, between the upper Mississippi, Lake Michigan, and Lake Superior. I say only the centre of this group; for I except the copper trap of the south side of Lake Superior, the age of which is much more recent, as well as the greater part of the dislocations which extend from Wisconsin River to Menomonee River. Most of the mountains in the northern part of New York State, between the Thousand Islands and Lake George, are also of the same epoch, as well as two little granitic groups in the State of Maine, between Moosehead Lake and St. John's River; also the mountains at the north of New Brunswick, near Bathurst, the Colequid Mountain, Nova Scotia, and, lastly, a small part of Cape Breton, between Louisburg and the Lake of Bras d'Or.

The three parallel lines of mountains, in Missouri, Arkansas and Texas, which I have named the Ozark system, appear to me also to belong to this older system of dislocation.

2. *Two Mountains and Montmorency System.* — The dislocations giving rise to this system took place at the end of the deposit of the first group of Lower Silurian; that is, after the formation of the Potsdam group. Its direction, according to the few observations that

I have been able to make, appears to be approximatively E. 40° N. and W. 40° S. The beds of the Potsdam group are greatly elevated near the junction with the metamorphic rocks, and are often metamorphosed themselves, which gives them a very hard, quartzite aspect. The localities where this system of dislocation can best be observed are the environs of Quebec, especially between the Montmorency and Indian Lorette Falls; Mount Calvary, in Two Mountains County, near Montreal, and Little Falls, in New York State. I think the dislocations may be regarded as of the same age that have taken place on the north-western side of the mountains at the north of the State of New York; and on some points of the line of junction of the Silurian and the metamorphic rocks which runs from Wisconsin to Menomonee River, in the eastern part of Wisconsin. The movement that gave rise to these dislocations was much less considerable than when the Lawrentine system appeared, and was not felt at distant localities; it has only modified some parts of the preceding upheaval, by crossing and penetrating it to form small chains adjacent to this older range of mountains.

3. *Montreal System.* — In many localities, and especially at Montmorency and at Little Falls, the beds of the second group of Lower Silurian, or Trenton group, are found deposited horizontally on very much inclined strata of the Potsdam group, presenting, consequently, very discordant stratification. These beds, forming the Trenton group, have also been subjected to dislocations soon after they were deposited. Without presenting any great projections, or marks of much disturbance and upheaval, these dislocations. which took place after the deposit of the second group of Lower Silurian, are, nevertheless, very clearly marked, and have left very remarkable traces, especially in Lower Canada. The summit of the mountain that overlooks Montreal is formed of dykes of greenstone or trap, which have entirely crossed the beds of the Trenton group, and are even spread over them. Several other dykes of trap which are found in the same position on different points of the borders of the Ottawa River, as well as the Mountains of Beloeil, Rougemont, Montanville and Johnson, near the Rivers Richelieu, Huron and Yamaska, appear to me to belong to the same system of dislocation, whose general direction would be precisely from east to west. I think traces of the

Montreal system will be found in other regions, and particularly in Upper Canada and the State of New York.

4. *Notre Dame Mountain System.*—If the preceding system has but slightly raised and disturbed the beds, it is not so with the present one, which I call the system of the Notre Dame Mountains, and which dates from the end of the deposition of the Lower Silurian. It will be remembered that the beds forming the third group of Lower Silurian are numerous strata of black schist, distinguished in the State of New York by the name of Utica and Hudson River group. These strata, which form almost entirely the banks of the River Richelieu, of the St. Lawrence lower than Montreal, and on which is situated the city of Quebec, have been upheaved along the whole of this line, to Cape Rozière at the extremity of Gaspé. The Notre Dame Mountains, formed of eruptive and metamorphic rocks, some of whose summits attain three thousand five hundred feet, owe their origin entirely to this movement, whose general direction appears to be E. 20° N., and W. 20° S. On looking at the map, it will be seen that the group of igneous rocks forming the Notre Dame Mountains is isolated, and entirely detached from the neighboring groups. A line of hills, of sedimentary rocks, of very slight elevation, extends between Madawaska and the River du Loup, and joins these mountains to those which are near Point Levi. Many of the geographical accidents of the chains which extend from Point Levi to Lake Champlain owe their origin, I think, to this system of dislocation; but the directions are uncertain, owing to the intercrossing with the system of dislocation which followed this, and formed the Green Mountains of Vermont. Nevertheless, considering only the Notre Dame Mountains, where the phenomena of dislocation are very decided, a well-marked and independent direction is easily recognized.

5. *Green Mountains, or the Meridional System.*—For a long time Mr. Hitchcock has indicated this system, which he distinguishes by the name of "Oldest Meridional and Hoosac Mountain System." Very much developed in the western part of Massachusetts, it forms entirely the Green Mountains of Vermont, and extends into Lower Canada as far as the River Chaudière. Its general direction approaches the meridian, with a slight deviation to the east, which gives for the average N. 7° E., and S. 7° W. The dislocations

which gave rise to this line of mountains took place much before the formation of the Alleghanies, yet, after the deposit of the Upper Silurian, as the metamorphic fossiliferous rocks that Mr. Logan has found in the Memphremagog and St. Francis Lakes prove. Thus I regard them as having appeared at the end of the Silurian period, and before the deposit of the Devonian beds. The igneous rocks, which then made an irruption through the surface of the earth's crust, upheaved very strongly all the sedimentary strata previously deposited; and, more, they have often overturned, folded and waved them, by submitting them simultaneously to a powerful metamorphic action, which was felt at a great distance from the eruptive centres. The Green Mountain system extends, as I have said, from the River Chaudière, Lower Canada, into Vermont, which it forms almost entirely; it comprehends Berkshire, and the line of hills that extends between the Connecticut River and Worcester, Massachusetts, Litchfield and Fairfield, Connecticut, and terminates in the environs of Bridgeport, at Long Island Sound. The boundary line between the States of New York and Connecticut presents crossings of this system with that of the Alleghanies, which much later has struck against the dislocations of the Meridional system, and even penetrated them in some places. The Geological Map of the State of New York, published by legislative authority, presents very clearly this point of the meeting of the two systems. Thus the valley of the Hudson, from Saratoga to West Point, is due to the Green Mountain system, and is parallel with its direction. The Green Mountains present, at several points of Vermont, and especially at the River Chaudière, Canada, quartzose veins traversing itacolumites, and containing native gold, which, though in some quantity, does not appear to present sufficient richness to reward the working of it.

The White Mountains probably owe a part of their elevation to this system. But the insufficiency of geological observations obliges me not to risk any supposition as to the age, or ages, of the chains of this group.

I consider as belonging to the Green Mountain system the elevation in Nova Scotia, between Merigomish, Arisaig and Cape St. George; as well as the line of dislocation which extends from the western extremity of the Gulf of Canso to Cape North, in the Island of Cape Breton. Finally, I think that the mountains extending from the Bay of Islands, at the western part of Newfoundland, to the Straits of Belle Isle, also owe their origin to this system of dislocation which took place at the end of the Silurian period.

6. Notwithstanding the impossibility of collecting certain and positive proofs of a system of dislocation which should have taken place at the end of the Devonian period, and before the Carboniferous deposits, I cannot refrain from noticing some indications, as I think, of disturbances in the deposits at this geological epoch. Thus, it seems to me, that, following the line of the Carboniferous formations in New Brunswick, and the great basin of the Ohio and Mississippi, there is a general disposition which gives to the strata a direction E. 8° S., and W. 8° N.; a direction which coincides with the systems No. 3 and 6, that Mr. Hitchcock has indicated in the south-eastern part of Massachusetts. These systems of Mr. Hitchcock have dislocated beds become metamorphic, which might easily be recognized as Devonian, and whose contours have formed the anthracite basin of Bristol County. It may happen, hereafter, that the metamorphic and eruptive rocks, indicated by Nicolet at the *Great Oasis*, and the pipestone quarries, Plateau du Coteau of the prairies, will be recognized as entering into this system; and, at all events, notwithstanding the extreme doubt with which I indicate this system No. 6, giving it no name of mountains to distinguish it, as I have done for the others, I none the less persist in regarding it as really existing, and as placed by its age between the dislocations of the Green Mountain and Alleghany systems.

7. *Alleghany System.* — Until now, geographers and geologists have called Alleghany or Appalachian chain all the mountains which extend from Montgomery, Alabama, to Cape Rozière and Gaspé, at the mouth of the St. Lawrence. According to the preceding classification, almost the whole of the mountains which form the States of Massachusetts, Vermont, New York and Maine, and Upper and Lower Canada, were much anterior to the apparition of the Alleghanies; and I have distinguished five or six systems of mountain ranges of different ages and directions, having no connection with the direction and age of the Alleghanies. For me, in a geographical point of view, the Alleghanies commence between Montgomery and Tuscaloosa, Alabama, and extend, without interruption, to the left bank of the lower part of Hudson River, in Putnam County, New

York. In a geological point of view, this system of Alleghanian dislocations continues into Connecticut to the eastward of New Haven, crosses Rhode Island, which it forms almost wholly, the eastern part of Massachusetts; passing by Lowell, it terminates at Portsmouth, New Hampshire; then it recommences from Cape Sable, Nova Scotia, and runs to Cape Canso, and forms all the eastern and central part of Newfoundland. The general direction of the Alleghany system is from north-east to south-west. A deviation more to the east is marked in the part which extends from the environs of Harrisburg, Pennsylvania, to the environs of New York city. This deviation arises from the meeting of the dislocations of this system with those of the Meridional system, which, in crossing and penetrating each other, have resulted in a mixed direction, at an angle of from eight to ten degrees further east, for the region of the Alleghanies between the Susquehanna and the Hudson. The influence of this deviation has extended much further to the south-west; and I do not hesitate to attribute to it very considerable faults, which are found in following the bases of the highest of the Alleghany Mountains in the western part of Pennsylvania, and Virginia, and in Tennessee, and which show, from time to time, sharp and angular curves in the form of bayonets.

The precise epoch of the dislocation of the Alleghany system is the end of the Carboniferous deposit. To this system is owing, almost wholly, the elevation of the Middle and Western States of the Union. A part of this elevation forms the magnificent ranges of mountains so celebrated as the Pennsylvania, Virginia and Kentucky Alleghanies; and that present an orographic regularity, of which explanations have been attempted that are far from satisfactory, when one is able to study the mountains themselves; and, we think, at present the phenomenon is too complicated, and the observations too few, to permit of general rules for the orography of this part of the Alleghany system.

Many parts of the Alleghanies, especially in Virginia, North Carolina and Georgia, contain auriferous itacolumites, which yield a profit to the miner, and have not been abandoned even since the discovery of the placers of California and Australia.

8. *Keewenau Point and Cape Blomidon System.* — The strata of the New Red Sandstone have also been subjected to dislocations that took place at the end of their deposit, and which, although they

have acted with but little regularity or intensity are none the less a special system, which is found very well characterized at Lake Superior and at the Bay of Fundy. After making many observations on the directions which these dislocations present, I have found an average direction of E. 35° N. and W. 35° S. In many localities, especially in the valleys of Connecticut and of New Jersey, where numerous dikes of trap are found that belong to this system, the direction is more northerly, and is confounded sometimes with that of the Alleghany system, and sometimes with the Green Mountain system; but that is owing, evidently, to crossings which have changed the primitive direction. To have the normal position of this system, it must be studied at Point Keewenaw, Isle Royale and Thunder Cape, Lake Superior; or on the two parallel shores of the Bay of Fundy, at Cape Split and Cape Blomidon, and also at the Magdalene Islands in the Gulf of the St. Lawrence. These dislocations surround immense dikes of basaltic trap, which has often flowed over them, covering the beds of the New Red Sandstone. This trap is chiefly remarkable for the veins of native copper which cross it perpendicularly, and give great value to the regions where this formation is found. Beside the native copper, all the varieties of copper ore, of native silver, and many zeolitic minerals, are frequent in this formation.

Dr. Charles T. Jackson first made known and characterized the age and the importance of the system of Point Keewenaw and Cape Blomidon. Until the publication of his report on the copper mines of Lake Superior, it was thought that the sandstone and the trap of Lake Superior were of the Lower Silurian epoch, or of the Old Red Sandstone; and it is this savant who, in connecting them with the trap and sandstone of Nova Scotia, New Brunswick and New Jersey, has placed them in their true stratigraphic position. Since this, Messrs. Whitney, Foster and Owen, have thought that the sandstone and copper trap of Lake Superior were indeed of the Silurian epoch; which was also the opinion of Messrs. Bayfield and Houghton, the first observers who described Lake Superior. The reader who is interested in the discussions on this subject will find in the bulletins of the Geological Society of France all the reasonings that have been made upon it: I will say here only that, having studied carefully these sandstones and copper traps in Nova Scotia, New Brunswick, Connecticut valley,

New Jersey, Maryland and Virginia, and having made the *complete tour* of Lake Superior, I do not hesitate to synchronize the whole, and, further, to assign to them the New Red Sandstone for their stratigraphic position, and for the age of dislocation the system following that of the Alleghanies. This opinion is not shaken by the last published reports of Messrs. Whitney, Foster and Owen. After having read them with the greatest attention, I have found no reason to change or modify my views. This system of copper trap forms Cape St. George, in Newfoundland, as well as the Magdalen Islands; and, according to the description of Mr. Logan, I think that several dykes of trap and beds of sandstone on the northern coast of the Bay of Chaleur, between the River Ristigouche, Richmond, and Port Daniel, are of the New Red Sandstone epoch, and of the Keewenaw Point and Cape Blomidon system.

I cannot point out with precision any system of direction of mountains following the dislocation of the New Red Sandstone. The Alleghanies, especially in the States of Virginia, North Carolina, Georgia and Alabama, have been subjected to new dislocations, which have either followed the primitive direction of this system or crossed it at a small angle. These dislocations probably took place at the end of the Jurassic and cretaceous epochs; also, perhaps, of the Lower Tertiary or Eocene.

As to the Rocky Mountains, the chains which form them are certainly to be ascribed to several systems of dislocation, and perhaps even little chains may be found there as ancient as the Silurian epoch. At least, it may be said now that several chains date from the epoch of the systems of the Green Mountains and the Alleghanies, and that the cretaceous, eocene and miocene formations, even, are very much upheaved, and consequently have been subjected to dislocations after their deposit.

If the golden placers of Lower Canada, Virginia, Georgia, California, Australia, the Ural Mountains and Eastern Brazil, are attentively considered, their position with relation to the sedimentary rocks that surround them, and the age of these sedimentary rocks, it will be seen that all these grand auriferous deposits are found in rocks whose dislocations have taken place during the second period of the palæozoic formations, from the end of the Silurian deposit to the Permian epoch.

M. de Beaumont, as I have said above, has endeavored to classify and connect with the European systems several mountain-chains of North America. Two of these attempts appear to be very exact, and to merit attention. One makes the Alleghany system coincide with the system of *Ballons des Vosges*, and the hills of Bocage, Brittany; this system, according to M. de Beaumont, which has dislocated the Carboniferous beds in Brittany, Westmoreland and the Vosges and Hartz Mountains, has, for the great circle of comparison, a great circle, passing by the top of the Brocken, and having a direction of W. 19° 15′ N. This great circle, affected by the curve of the earth, passes along the shores of the Atlantic, and when it reaches Washington its direction is E. 46° 55′ N. and W. 46° 55′ S. That is, it coincides exactly with the direction of the Alleghany system, since we found the mean direction of this system to be from north-east to south-west; namely, N. 45 E. and S. 45° W.

The second system of Europe, which is connected with those of America by M. de Beaumont, is the system of the Thuringerwald and of the Morvan, which he unites with our Keewenaw and Cape Blomidon system.

The system of the Thuringerwald has dislocated in Europe the trias or New Red Sandstone over a large extent of country, mainly in Thuringia, Bohemia, the Vosges, the Morvan, near Dublin, and at Trees, in Shropshire, Ireland. The great circle of comparison for this system has been carried by the top of the Greifenberg, Thuringerwald, and directed at this point W. 39° N. and E. 39° S. By prolonging this great circle to America, it will be seen to pass very near Point Keewenaw, and to be parallel to the central axis of this point, as well as to the axis of Isle Royale, and to the circle which might be drawn from Annapolis to Cape Blomidon, Nova Scotia.

To complete the analogy which exists between these systems of the Thuringerwald and of Point Keewenaw, I will mention that Mr. Murchison has pointed out in Ireland, at the cliffs of Portrush, basaltiform trap of this epoch; and, more, this trap is cut perpendicularly by copper veins, exactly like those in the Bay of Fundy, New Jersey, and at Lake Superior.

M. de Beaumont also tries to prolong in America his system of Morbihan, which he regards as having taken place before the deposit of the Silurian formation. This prolongation is very far from presenting the exactness of the two preceding ones; and I think it very problematic, at least with the great circle of comparison given to it by M. de Beaumont. It is true that these great circles of comparison are only temporary; and, perhaps, when it is modified by later observations, this system may coincide with the Lawrentine, or rather with the Two Mountains and Montmorency system.

In concluding this chapter on the systems of mountains, I must repeat that it is a very imperfect attempt, and that I give it only as a first essay, which may become hereafter an exact and rigorous classification of the different mountain ranges which are found in the United States. At least, the relations which I think are to be recognized between the seven first systems and the seven principal groups of American palæozoic rocks are facts which merit attention; and I hope they will be acknowledged as wholly or for the most part true by the geologists who study North America.

CONCLUSION.

In this rapid sketch of the geological constitution of North America, it has been shown how profusely the most useful minerals, and even the precious metals, are scattered over the various parts of this vast continent. In striking contrast to the European, who finds himself in a locality very much circumscribed by natural limits, and also by political divisions, the American sees spread out before him a mineral wealth almost without rival, and which covers so vast an extension of country that it cannot be monopolized or concentrated in the hands of a few individuals; and thus it may be said, with truth, that the mineral kingdom itself contributes to the development of American democracy. It is well known that the agriculture and the vegetation are more democratic,—that is to say, more uniform,—in the United States than anywhere else; and the words of my fellow-countryman De Tocqueville are confirmed, when he says, in the "*Democracy in America*," "In the United States it is not only the legislation that is democratic; nature herself works for the people."

How can we but be astonished at the sight of the development of America? All things combine to aggrandize her power: the distribution of rivers, lakes, mountains, plateaus and plains; the agricultural system which results from this as an immediate consequence; and even the strata, and the rocks themselves, enclosing in their bosoms treasures for industry, manufactures, naval enterprise and even for exportation,— and on an immense scale, as is everything in this country. In view of these facts, we may say now that the winds that swelled the sails of the *Mayflower*, and guided its prow to the rocks of Plymouth, wore a mark of the protection of Divine Providence for the Pilgrims; and that, in giving North America to the Puritans of England, the Almighty Governor of the world gave the most propitious land for the development of the political and religious ideas they had embraced, and which they and their descendants have continued to propagate with so much success.

APPENDIX.

A.

GENERAL VIEW OF THE GEOLOGICAL STRUCTURE OF THE WORLD.

We give below the introduction to Mr. De La Bêche's excellent work — *The Geological Observer*. The only alteration is in the table of *Upper Stratified or Fossiliferous Rocks*, which is very much enlarged, and the suppression of the notes. *The Geological Observer* is, without doubt, the best elementary treatise which has appeared in the English language; and we cannot recommend it too highly to those who would thoroughly understand the best methods of studying and observing in geology.

Observations have now been sufficiently extended and multiplied to show that, during a long lapse of time, the surface of our planet has been undergoing modifications and changes. Of these the most marked have been produced by the uprise of mineral matter, in a molten state, from beneath that surface; the wearing away and removal to other localities of this matter, either in its first state after cooling, or in some secondary condition, by atmospheric influences and waters variously distributed for the time being; the preservation of the remains of animal and vegetable life, during at least a portion of this lapse of time, amid deposits accumulated, for the most part, in horizontal layers beneath waters, and by the unquiet state of the earth's surface itself, from which, while considerable areas have been at different times raised slowly above and depressed beneath the level of the ocean, occasionally whole masses of mineral matter of various kinds have been squeezed, bent and plicated, sometimes ridged up into ranges of mountains.

To enable the geologist systematically to proceed with his researches, it became as needful for him as for other cultivators of science to have the power of classifying his observations. Of the various classifications proposed or modified at different times, to satisfy the amount of knowledge of those times, it would be out of place here to make mention, further than to remark that at present a more mixed classification is often employed than seems desirable; — for example, it is not unusual for the term *tertiary*, or *tertiaries*, to be applied to all accumulations posterior to the chalk of the western Europe, while the other terms, of secondary and primary or primitive, to which it has reference, are scarcely or seldom now mentioned. We have, again, a mixed nomenclature for the groups of deposits, or the deposits themselves, for which it has been thought desirable to find distinctive names. While some groups are referred to localities, such as Silurian, Jurassic, Neocomian, and the like, others are named after some circumstance supposed characteristic; such as Carboniferous, from containing the great coal deposits of Europe and North America; or, Oölitic, from many of the limestones in it being oölitic, that is, resembling the roe of a fish, being composed of numerous small, rounded grains, formed of concentrically arranged coatings of calcareous matter.

It has been often considered that names derived from localities where certain deposits have been taken as types are preferable to those pointing to any mineral structure, inasmuch as not only can the geologist readily make himself familiar with the kind of accumulations intended to be represented by the names, by visiting and studying the localities whence they are taken, but as also particular mineral structures having been repeated as often as the conditions for them arose, they form no guide for determining the relative age of rocks, whatever may have been the impression when names of that kind were given, and geological science less advanced than at present. The two structural names mentioned are thus liable to objection, carboniferous deposits extending from an earlier period than that supposed to be represented by the term, and up to the higher accumulations above the cretaceous series inclusive, and the oölitic character reaching from limestones amid the earlier fossiliferous rocks to the present day.* The mixed character of the present geological nomenclature arises, no doubt, from the manner in which, from time to time, various geologists have directed attention to different rocks or accumulations of them; those names having generally remained which have been found convenient and sufficient, up to the present time, for the purposes for which they have been employed.

The igneous products being those from which the chief part, if not the whole, of the detrital and even chemical deposits have been directly or indirectly derived, it would appear desirable to consider them in the first place. Whatever the views entertained of the fluid condition of our planet, whence its form has resulted, such fluid condition produced by heat sufficient to keep all its component parts in that state, the present condition of the earth's surface in dispersed localities shows an abundance of points through which igneous products are now ejected; and the more extended the observation, the more certain does the inference appear correct, that the like has happened from the earliest times, —at least, since the seas were tenanted by life. It has also been ascertained that molten matter has risen from beneath in more massive forms, and in a manner

* One of the limestones of the Lower Silurian series in North Wales, the Rhiwlas, near Bala, is oölitic. The Lower Carboniferous presents beds of oölitic limestone in Kentucky at the Mammoth Cave, and in Tennessee. The modern rocks of the Florida coasts also contain oölitic limestone, exactly like the coral rag oölite of England and France.

with which we are not familiar as now occurring, though such molten masses may, indeed, be formed at depths in the earth's crust whence only future geological changes could bring them above the level of the sea. At all events, this massive form of intrusion is found amid comparatively recent geological accumulations, as well as among those of the most ancient date.

The mode of occurrence of the igneous rocks would seem to point to their classification according to their chemical and mineralogical characters, so that any resemblance or difference, that may exist between them, may be traced through the lapse of geological time, the relative dates of their appearance being obtained by means of the accumulations with which they may be associated, and to which relative geological dates can be assigned. The following sketch of the more prominent of the igneous rocks may here suffice:

Granitic Rocks.—Those composed of a granular mixture of quartz, felspar (whether orthoclase, albite or labradorite) and mica, with occasionally the addition of schorl, and some other minerals. As the aspect of these rocks varies considerably, according to original chemical composition or the mode of cooling, a great variety of appearances are assumed, to which names have been assigned. It thus becomes desirable that these characters should be given whenever it can be accomplished, and that the mere term *granitic* be accompanied by mineralogical detail, and by a statement of the chemical composition, so that correct data may be collected for a proper appreciation of the real differences and resemblances of the rocks commonly thus named.

Felspathic Rocks.—The separation of these from the foregoing may often be regarded as somewhat imaginary; as, indeed, is the case with definite classifications of the great bulk of the igneous rocks, passing, as they sometimes do, into each other in masses of no very extraordinary volume. The variety known as compact felspar is most frequently a compound of the elements of some felspar with a surplusage of silicic acid beyond that required for the silicates of that mineral; so that, when opportunities have occurred for crystallization of the parts, the result has been a compound of felspar and quartz, or a *granitello*, as it has been sometimes termed,—in that case, a modification of the granitic rocks, when the same minerals may also constitute a portion of a general mass. The *trachytes* of active volcanoes, and those termed extinct and of comparatively recent geological date, may represent the more pure felspathic rocks, when wholly formed of felspars, though it would appear that similar rocks are also found amid the igneous products of very ancient geological periods. Felspathic matter—that is, the various component substances in proportions which would form minerals of the felspar family (allowing for that substitution of one substance for another, termed *isomorphism*)—if crystallized, should at least constitute the great bulk of these rocks, whatever others may be entangled among them.

Hornblendic Rocks.—These, including among them the rocks in which augite is substituted for hornblende, form a somewhat natural division, so far as the prevalence of these minerals may be sufficient to give a character to the mass of an igneous rock; inasmuch as silicate of lime is a marked ingredient, in addition to the silicate of magnesia, another essential substance, and protoxide of iron, generally present, sometimes replacing much of the lime and magnesia. In this division, therefore, are included the dolerites and basalts of active and extinct volcanic products, and the greenstones, generally of more ancient date. In dolerites, silicate of lime is also present in the labradorite, when that member of the felspar family is mingled with the augite of that rock. Taken as a whole, the hornblendic or augitic rocks are compounds of those minerals and some member of the felspar family, there being sometimes an excess of silica beyond the amount required for the various silicates in the hornblende or augite, and felspar; this excess, then, as it were, thrust aside as quartz.

Serpentinous Rocks.—To a certain extent these also appear a somewhat natural group of igneous products, especially when viewed with reference to a peculiar aspect, and the presence of silicate of magnesia and combined water, as constituting the bulk of the rock. The rocks of this division vary, however, somewhat materially in their constituent substances, and in the proportions of them. Taking *bronzite* to be the mineral usually named diallage, it would appear little else than the silicate of magnesia of the matter of the purer serpentine, mingled with a minor proportion of protoxide of iron, and a little alumina crystallized, a small quantity of water also forming a part of it. The mineral now chiefly named *diallage* contains sufficient lime, in addition, to make it essentially a silicate of lime and magnesia, with also a marked quantity of oxide of iron. In the compound, sometimes largely crystallized, termed diallage rock (gabbro), and not unfrequently associated with serpentine, the so termed diallage has to be carefully examined. In all these rocks, whatever their variations, magnesia is a marked ingredient.

Porphyritic Rocks.—Though, no doubt, various kinds of mineral matter which have been in a molten state may be porphyritic,—that is, have some mineral or minerals crystallized out and apart from the mass of the remainder of the rock,—it seems nevertheless convenient, for the present, to notice those rocks as a group. Even amid vitreous matter, from comparatively quick cooling after fusion, definite chemical combinations may be crystallized, and dispersed through such matter. This can be artificially accomplished in our laboratories, and silicate of lime in crystals can be obtained, dispersed through ordinary glass. In the arrangement of particles, beyond the vitreous condition, forming the compact and stony state, the porphyritic character is not rare among rocks; crystals, such as those of felspar, being dispersed amid a base of compact mineral matter. When the latter is chiefly felspathic, the rock is usually known as *felspar porphyry.* In like manner crystals of other minerals are also thus dispersed amid a similar base, such as those of quartz and mica. The base or general mass of the rock is occasionally granular, such as a compound of felspar and hornblende, constituting greenstone, with dispersed crystals of felspar or hornblende, such base having thus advanced to a state of confused crystallization. These are usually termed *greenstone porphyries.* In like manner, certain granites become porphyritic, from separate crystals of felspar being scattered among the general compound, confusedly crystallized, and the rock is then called a *porphyritic granite.* Even serpentine becomes in a manner porphyritic, when crystals of bronzite or diallage are dispersed through a base of that rock. The apparent conditions are, that the chemical composition and the mode of cooling of the general mass are such that certain constituent substances can combine and form separate and definite crystallized bodies, the remainder of the rock either not attaining the state when definite mineral compounds can be found, or only doing so after the production of the first formed minerals, and then in a confused manner, not interfering with the forms of the crystals first produced.

With regard to the mineral accumulations derived either directly or indirectly from the igneous rocks, and spread over areas of varied extent and form, by means of water, there is a large mass, more or less characterized by the presence among it of the remains of animals and plants existing at different periods, and so perishing that portions of them, commonly only the harder parts, have been entombed in the mineral accumulations of such different times.

Observation has shown that these accumulations have succeeded one another, as the various detrital deposits in lakes and seas now succeed those which have preceded them, so that when the ancient sea or lake bottoms, which, elevated into the atmosphere, now constitute so large a portion of dry land, can be studied in cliffs or other natural sections, or by artificial cuttings or perforations, their manner of succession can be ascertained. The more investigations have advanced, the more does it appear that these organic-remain bearing, or *fossiliferous rocks*, as they have been termed, have been deposited and arranged as similar accumulations now are in rivers, estuaries, lakes and seas. Hence, the geologist, in endeavoring to ascertain the range of such fossiliferous deposits at any given time upon the earth's surface, has to consider the relative amount and position of the land and waters of that time, with all their modifying influences, as also the various conditions under which the life of the period may have been distributed, and its remains entombed amid the detrital and chemical deposits of the day. In fact, he has, from all the evidence he can collect, to suppose himself studying the state of the earth's surface, at such given time, as well with respect to its physical condition as the existence and distribution of life upon it.

Viewing the fossiliferous rocks in this manner, it may be that some of those divisions among them, which it has been found convenient to make for their more ready description, and the tracing of certain states of a sea-bottom over minor areas, have been too minute, — regarded as divisions applicable to the surface of the earth generally, since it is not to be supposed that particular mud or sand banks, however considerable locally, were more likely to have been formerly continued, even at intervals, over the earth's surface, than they now are. At the same time, such minor divisions showing the constancy or modification of conditions, as the case may be, over the minor areas, are important, inasmuch as it is by a correct appreciation of this detail, and the careful consideration of how much may be regarded in that light, and how much as more general, that we learn the true value of the latter, and the restrictions which should be placed upon our views derived from the former.

Assuming the general condition of the earth's surface during the accumulation of the varied deposits in which the remains of animal and vegetable life have been entombed to have been formerly much as at present, — regarding the subject on the large scale, and without reference, for the moment, to the variable distribution of land and water, or to whether the heat in the earth itself may or may not, in remote times, have had a greater influence on the life of those times than at present, — the sea would appear to have been the chief receptacle of the various mineral accumulations of all periods; so that classifications of the fossiliferous rocks, founded on a succession of deposits in it, would probably be alike the most useful and natural. The manner in which marine invertebrate animals now live, and the mode in which the remains of similar animals occur amid the fossiliferous rocks, is such that this division of life seems now very generally admitted as the most appropriate on which to base classifications founded on the distribution of animals the remains of which are discovered entombed in rocks. It will be sufficient here to mention that, after duly first ascertaining the actual relative superposition of the various mineral accumulations themselves, for evidence of their real succession, and examining the remains of animal and vegetable life which have been found in them, it has been inferred that certain minor and major divisions may be effected in the general mass which shall represent the kinds of sea-bottoms making given and succeeding geological times. Without, in the least, doubting that much modification may not be found needed in classifications founded upon the examinations of even considerable areas, when an effective classification, representing the main facts connected with the accumulation and spread of fossiliferous rocks over large portions of the earth's surface, may be needed, it still becomes desirable to have that which may satisfy the requirements for the time being. The following sketch, therefore, of the general divisions at present considered desirable for the area of Europe, and supposed in part, at least, to be found also convenient for the mode of viewing the fossiliferous deposits in many other parts of the world, may be useful, especially as respects the major divisions.

UPPER STRATIFIED OR FOSSILIFEROUS ROCKS.

I. Modern.
II. Quaternary.
III. Tertiary.
IV. Secondary.
V. Paleozoic.*

I. Modern Rocks.

Comprising all mineral accumulations of the present time, namely, Alluvial of the rivers, Coral islands and reefs, Thermal springs, Glacier moraines, Icebergs, Tide deposits, etc.

* The name *Paleozoic* has replaced advantageously that of *Primary*, which had an equivocal meaning, and was often employed to designate eruptive rocks, or metamorphic schists, often of a more recent origin than the Paleozoic rocks. Geologists, in search of classification, took advantage of the opportunity to make new names in *ic*, such as *Azoic*, *Hypozoic*, *Mesozoic*, *Cainozoic*, etc., which have not been adopted, as they were not so appropriate as the names they were intended to replace.

Further, the name of *Azoic* would not apply to a special group of rocks, for azoic strata are found in all the formations. It has happened, in consequence, that the geologists who have used this term have placed in the *Azoic group* rocks of very different ages.

II. Quaternary.

Comprising the Diluvial rocks, the Drift and Erratic blocks, Cavern deposits and Osseous Breccias, Loess of the Rhine, etc.

III. Tertiary.

A. Upper Tertiary Group.	Pliocene. Crags of Suffolk, England. Antwerp Crag; Sands of Diest, Belgium. Cotentin Crag; Faluns de la Loire; Sables de la Sologne et des Landes, France. Oeningen deposit, Switzerland. Subapennine beds, or Mattajoue; Creta di Sienna, Italy. Aralo Caspian, or Steppe Limestone, Russia.
B. Middle Tertiary Group.	Miocene. Wanting in England. Bolderberg sands, Belgium. Calcaire à Helix de la Beauce; Fontainebleau Sandstone; Molasse de Leognan et Calcair de Bourg near Bordeaux; Sansan deposit; lower beds of la Limagne d'Auvergne; Gypse d'Aix, France. Nagelfluh and Molasse, of Switzerland. Molasse of Superga and Bormida's valley; Coal of Caddibuona, Piedmont. Salt deposits of Wieliczka, Poland.
C. Lower Tertiary Group.	Eocene. Barton Clay; Bagshot Sand; London Clay. Limburg beds; Brussels beds. Gypse de Montmartre; Grès de Beauchamp; Calcaire Grossier; Argile plastique de Paris; Molasse du Fronsadais et Calcaire de Blaye près de Bordeaux. Nummulitic formation. Dax, Biaritz and the Corbières deposits. Flysch of the Alps. Nummulitic Sandstone of Kressenberg, Austria. Ronca and Monte Bolca deposits, Vicentine. Macigno and Alberese of the Apennines. Kief and Antipofka rocks, Russia.

IV. Secondary.

A. Cretaceous Group.	a. Chalk of Maestricht and Valkenberg, Holland; Limestone of Faxoe, Denmark. b. White Chalk, or Chalk with flints, England; Craie de Meudon, France; Marly Chalk of Lusberg, Holland; Hippurite Limestone of Italy; Chalk of Kursk and of the country of the Don Cosacks. c. Lower Chalk, Chalk-marl and upper Greensand of England; Craie Tuffeau de France; Dieves and

A. Cretaceous Group.

Tourtia of Hainaut, Belgium; Lower Chalk of Aix-la-Chapelle, Rhenish Prussia; white and red Scaglia of Italy; Carpathian Sandstone of Cracow and Orlova, Gallicia.

D. Gault of England and France; Grey Scaglia, Italy; Quadersandstein of Saxony; Wienner Sandstein.

E. Lower Greensand of England; Neocomian of Switzerland, France, Crimæa and Caucasus; Biancone of Italy.

B. Jurassic Group.

A. Wealden. Weald Clay, Hastings Sand and Purbeck beds of England; Turtle Limestone of Soleure, Switzerland; Wealdens of Hanover and Westphalia.

B. Upper Oolite. Portland beds, Kimmeridge Clay and Coral Rag of England; Portlandian, Kimmeridgian, Sequanian and Corallian formations of France and Switzerland; Solenhofen Schists and Nattheim Coral Schists, or White Jura of Swabia and Franconia; Majolica of Italy.

C. Oxfordian. Oxford Clay, with Kelloways Rock, England; Argovian, Dives Clays and Kellovian of France and Switzerland; Spongites beds, Ornati Clay and Macrocephali bed of Swabia; Upper Alpine Limestone; Limestone of Porte dé France of Grenoble, Dauphiné; Ammonitico Rosso of Italy; Moskwa beds of Russia.

D. Lower Oolite. Cornbrash, Forest Marble, Bradford Clay, Bath Oolite, Fuller's Earth, Inferior Oolite of England; Vesulian Marls, Coral Limestone, Ladonian Limestone and Bayeux Oolite of France; Brown Jura of Swabia and Franconia.

E. Lias. Whitby Shale, Lyme Regis Shale, Marlstone, Downcliff Sandy Marl, Aberthan Blue Marl and Linksfield Sandstone of England; Superliassic Grit, Vassy Cement bed, Middle Lias, Gryphæa arcuata Limestone of France and Switzerland; Jurensis Marl, Boll Schists, Balingen Clays, Arietes and Cardinia beds, or Black Jura of Swabia; Lower Alpine Limestone.

C. Trias Group.

A. Keuper. Yellow and White Sandstones, Grellfarbige Mergel, Schilfsanstein, Gyps and Mergel, Lettenkohle of Wurtemberg; Variegated Marls of England; Marnes Trisées of France; Verrucano of Italy.

B. Muschelkalk. Hauptmuschelkalk, Salzgebirge, Wellenkalk, Wellendolomit of Wurtemberg; Calcaire de Lunéville, France; St. Cassian beds, Tyrolese Alps; Limestone of Recoaro, Venetian Alps.

C. TRIAS GROUP.
- C. Bunter Sandstein. Thonige and Kieselige Sandsteine, and Rothe Letten of Wurtemberg; Grès Bigarré of France; Upper New Red Sandstone of England.

D. PERMIAN GROUP.
- A. Magnesian Limestone of England; Zechstein and Copper Schists of Thuringia; Upper Permian of Russia.
- B. Lower New Red Sandstone of England; Vosgian Sandstone of France; Todtliegendes of Germany; Lower Permian of Russia.

V. PALÆOZOIC.

A. CARBONIFEROUS GROUP.
- A. Coal Measures, Millstone grit of England; Terrain Houiller de France; Kohlengebirge of Germany; Goniatite grits of Russia.
- B. Carboniferous or Mountain Limestone of England; Terrain Anthracifère de la Sarthe, France; Bergkalk of Germany; Fusulina Limestone, Moscow Limestone, Lower Limestone of Tula and Kaluga, Russia.

B. DEVONIAN GROUP.
- A. Upper Old Red Sandstone, or Petherwin Slate and Marwood Sandstone of Scotland and Wales; Grundt Limestone of the Hartz; Domanik Schiefer of Timans Mountains, Russia.
- B. Slate Rock and Limestone of South Devonshire, Plymouth Limestone, the Asterolepsis formation of Stromness, Caithness and Orkney Sandstone, England and Scotland; Bas-Boulonnais beds, Limestone of Sablé, France; Limestone of Sabero, Leon, in Spain; Limestone of the Eifel, Germany; Riga and Dorpat beds, Russia.

C. UPPER SILURIAN GROUP.
- A. Upper Ludlow, Aymestry Limestone and Lower Ludlow of England; Flagstones and Tilestones of Steens fiord, Norway; Upper Limestone and Schists of Bohemia.
- B. Wenlock Limestone, Wenlock Shale and Woolhope Limestone of England; Coralline Limestone of Christiania; Isles of Gottland and Oesel in the Baltic; Schists and Red Sandstone of St. Jean sur Erve, France; Middle and Lower Limestones of Bohemia.

D. LOWER SILURIAN GROUP.
- A. Caradoc Sandstone of England; Graptolites Schists of Sweden and France; Pentamerus Limestone of Jelebeck, Norway.
- B. Llandeilo and Builth flags of England; Ardoises à Trilobites d'Angers, France; Etage des Quartzites of Bohemia; Pleta or Orthoceratites Limestone of Scandinavia and Russia.

D. Lower Silurian Group.	c. Bala beds, Tremadoc Slates and Lingula Flags of England; Drammen Flags and Sandstone of Christiania, Norway; Protozoic Green Schists of Bohemia; Ungulite Grit and Blue Shale of St. Petersburg, Russia.

LOWER STRATIFIED ROCKS.

Although alteration in the mineral character of the fossiliferous rocks, from the influence of intruded igneous matter in a molten state, or arising from other modifying causes, often produces mica slates, hornblende slates, gneiss, and other forms of laminated and stratified deposits with a peculiar aspect, there appears, nevertheless, evidence in Scandinavia and the British islands, and also in other parts of Europe, to show that beneath all the fossiliferous rocks above noticed there are mica and chlorite slates, quartz rocks, crystalline limestones, gneiss, hornblende, and other rocks of earlier production. These may, indeed, be merely altered or metamorphosed detrital and chemical deposits of earlier times, and possibly organic remains may be eventually discovered in them; but, until this shall happen, it seems desirable to keep them asunder, for the convenience of showing previous accumulations to those noticed in the last division.

It would be out of place to attempt extended descriptions of the various rocks noticed in the above sketch. Information respecting them will be obtained by reference to works in which such descriptions are inserted, and still better by studying collections, with the aid of a competent person, in which their varied characteristics, as well mineral as paleontological (when fossiliferous), may be carefully considered and effectively displayed. The field, however, is the great source of geological knowledge, however important the cabinet, in its place, may also be; — it is there that the observer learns to appreciate the greater problems of geology, and where he may himself so materially assist in obtaining correct views of the modifications which the earth's surface has undergone in past times, and of the causes tending to obliterate its present condition.

B.

LIST OF BOOKS CONSULTED FOR THE COMPLETION OF THIS WORK.

A GREAT number of memoirs on the Geology of North America having been published in the scientific journals, I refer the reader to these collections; in their tables of contents are the titles of these memoirs, and the names of their authors. I will name, however, several geologists whose works, published in these reviews, have been of great service to me. They are, Messrs. Baddeley, Magdalen Islands; Bayfield, Newfoundland, Anticosti and Lake Superior; Brown and Dawson, Cape Breton and Nova Scotia; Safford, Tennessee; Rœmer, Texas; and H. King, Missouri.

The American Journal of Science and Arts; conducted by B. Silliman and B. Silliman, Jr., and James D. Dana. First and second series. New Haven.

Transactions of the Literary and Historical Society of Quebec. Quebec, 1829 to 1843.

Reports of the Association of American Geologists and Naturalists; 1840, 1841, 1842, 1844.

Proceedings of the American Association for the Advancement of Science; 1847 to 1851.

The Quarterly Journal of the Geological Society of London. London, 1844 to 1852.

Bulletin de la Société Géologique de France. Paris, 1830 to 1852.

UNITED STATES AND BRITISH AMERICA.

Observations on the Geology of the United States of America. By William Maclure. Philadelphia, 1817.

Travels in North America in the years 1841—42, *with Geological Observations on the United States, Canada and Nova Scotia.* By Charles Lyell. In two volumes. New York, 1845.

A Second Visit to the United States of North America. By Sir Charles Lyell. In two vols. New York, 1849.

A Geological Nomenclature for North America. By Amos Eaton. Albany, 1828.

Statistics of Coal. By Richard Cowling Taylor. Philadelphia, 1848.

Notice sur les Système de Montagnes. Par L. Elie de Beaumont. Paris, 1852.

The Taconic System. By Ebenezer Emmons. Albany, 1844.

Monograph of the Fossil Squalidæ of the United States. By Robert W. Gibbes. Philadelphia, 1848 and 1849.

Observations on the Geology and Organic Remains of the Secondary, Tertiary and Alluvial Formations of the Atlantic coast of the U. S. of North America. By Lardner Vanuxem and S. G. Morton. Philadelphia, 1828.

Description of some new species of Organic Remains of the Cretaceous group of the U. S. By Samuel George Morton. Philadelphia, 1842.

Contributions to Geology. By Isaac Lea. Philadelphia, 1823.

Synopsis of the Organic Remains of the Cretaceous Group of the U. S. By Samuel George Morton. Philadelphia, 1834.

Lake Superior: its Physical Character, Vegetation and Animals. By L. Agassiz. Boston, 1850.

BRITISH PROVINCES AND HUDSON'S BAY TERRITORY.

Journey to the Shores of the Polar Sea, in 1825—1827. By Captain John Franklin. London, 1829.

Narrative of the Arctic Land Expedition. By Capt. Back. Paris, 1836.

Arctic Searching Expedition: a Journal of a Boat-voyage through Rupert's Land and the Arctic Sea, in search of the Discovery Ships under command of Sir John Franklin. By Sir John Richardson. In two vols. London, 1851.

Geological Survey of Canada; Reports of Progress for the years 1842 to 1852. By W. E. Logan, Alex. Murray and T. S. Hunt. Montreal.

Newfoundland in 1842. By Sir Richard Henry Bonnycastle. In two volumes. London, 1842.

Excursions in and about Newfoundland during the years 1839 *and* 1840. By J. B. Jukes. London, 1842.

Reports on the Coal Fields of Carribou Cove and River Inhabitants, Cape Breton. By J. W. Dawson. Pictou, 1848.

Remarks on the Mineralogy and Geology of Nova Scotia. By Charles T. Jackson and Francis Alger. Cambridge, 1832.

Report on the Albert Coal Mine (New Brunswick). By Charles T. Jackson. New York, 1851.

Remarks on the Geology and Mineralogy of Nova Scotia. By Abraham Gesner. Halifax, 1836.

Geological Survey of the Province of New Brunswick; Report of Progress for the years 1839 to 1843. By Abraham Gesner. St. John's.

New Brunswick. By Abraham Gesner. London, 1847.

NEW ENGLAND STATES.

Geological Survey of the State of Maine; Reports of Progress for the years 1837 to 1840. By Charles T. Jackson. Augusta.

First Annual Report on the Geology of the State of New Hampshire. By Charles T. Jackson. Concord, 1841.

Views and Map, illustrative of the Scenery and Geology of the State of New Hampshire. by Charles T. Jackson. Boston, 1845.

Geography and Geology of Vermont. By Zadock Thompson. Burlington, 1848.

Annual Report on the Geology of the State of Vermont; for the years 1845 to 1848. By C. B. Adams. Burlington.

Final Report on the Geology of Massachusetts. By Edward Hitchcock. Amherst, 1841.

Report on the Geological and Agricultural Survey of the State of Rhode Island. By Charles T. Jackson. Providence, 1840.

Report on the Geology of the State of Connecticut. By James G. Percival. New Haven, 1842.

A Report on the Geological Survey of Connecticut. By Charles Upham Shepard. New Haven, 1837.

MIDDLE STATES.

Geology of New York. In four volumes. By Matter, Emmons, Vanuxem and Hall. Albany, 1843.

Paleontology of New York; vols. I. and II. By James Hall. Albany, 1847 and 1852.

A Geological and Agricultural Survey of the District adjoining the Erie Canal, in the State of New York. By Amos Eaton. Albany, 1824.

Description of the Geology of the State of New Jersey. By Henry D. Rogers. Philadelphia, 1840.

Geological Survey of the State of Pennsylvania; Report of Progress for the years 1836, 1838 to 1842. By Henry D. Rogers. Harrisburg.

Memoir of the Geological Survey of the State of Delaware. By James C. Booth. Dover, 1841.

Annual Report of the Geologist of Maryland. By J. T. Ducatel. Baltimore, 1840.

Geological Survey of the State of Virginia; Report of Progress for the years 1836 to 1840. By William B. Rogers. Richmond.

WESTERN STATES.

Geological Survey of the State of Ohio; Report of Progress for the years 1837 and 1838. By W. W. Mather. Columbus.

Contributions to the Geology of Kentucky. By Lunsford P. Yandell and B. F. Shumard. Louisville, 1847.

Geological Survey of the State of Tennessee; Report of Progress for the years 1836 to 1841, and 1844, 1845, 1848. By G. Troost. Nashville.

Geological Survey of the State of Indiana; Report of Progress for the years 1837 and 1838. By David Dale Owen. Indianapolis.

Geological Survey of the State of Michigan; Report of Progress for the years 1837 to 1840. By Douglass Houghton. Detroit.

Report on the Geological Survey of the Mineral Lands of the United States in the State of Michigan. By Charles T. Jackson. Washington, 1850.

Report on the Geology and Topography of a Portion of the Lake Superior Land District in the State of Michigan. By J. W. Foster and J. D. Whitney. In two parts. Washington, 1850 and 1851.

Report of a Geological Exploration of part of Iowa, Wisconsin and Illinois. By David Dale Owen. Washington, 1844.

Report of a Geological Reconnoissance of the Chippewa Land District of Wisconsin, and of a part of Iowa and Minnesota. By David Dale Owen. Washington, 1849.

Report of a Geological Survey of Wisconsin, Iowa and Minnesota, and incidentally of a portion of Nebraska Territory. By David Dale Owen. Philadelphia, 1852.

Report of a Geological Reconnoissance made in 1835, *from the seat of Government, by the way of Green Bay and the Wisconsin Territory, to the Coteau de Prairie.* By G. W. Featherstonhaugh. Washington, 1836.

Geological Report of an Examination made in 1834 *of the Elevated Country between the Missouri and Red Rivers.* By G. W. Featherstonhaugh. Washington, 1835.

SOUTHERN STATES.

Report on the Geology of South Carolina. By M. Tuomey. Columbia, 1848.

Statistics of the State of Georgia. By George White. Savannah, 1849.

TERRITORIES, INDIAN COUNTRY, ROCKY MOUNTAIN AND CALIFORNIA.

Report intended to Illustrate a Map of the Hydrographical Basin of the Upper Mississippi River. By T. N. Nicollet. Washington, 1843.

Report of the Exploring Expedition to the Rocky Mountains in the year 1842, *and to Oregon and North California in the years* 1843—44. By Captain J. C. Frémont. Washington, 1845.

Geographical Memoir upon Upper California. By Captain J. C. Frémont. Washington, 1848.

Notes of a Military Reconnoissance from Fort Leavenworth, in Missouri, to San Diego, in California; made in 1846—47, with the Advanced Guard of the "Army of the West." By Major W. H. Emory. Washington, 1848.

Report of an Examination of New Mexico in the years 1846—47. By Lieut. J. W. Abert. Washington, 1848.

Memoir of a Tour to Northern Mexico, connected with Colonel Doniphan's Expedition, in 1846—47. By A. Wislizenus. Washington, 1848.

Journal of Captain A. R. Johnston, from Santa Fé to San Diego. Washington, 1848.

Report and Map of the Route from Fort Smith (Arkansas) to Santa Fé (New Mexico). By Lieut. J. H. Simpson. Washington, 1849.

Geology of the United States Exploring Expedition, under the command of Charles Wilkes. By James D. Dana. New York, 1850.

Reports of the Secretary of War, with Reconnoissances in New Mexico, Texas, &c. Washington, 1850.

Exploration and Survey of the Valley of the Great Salt Lake of Utah. By Capt. Howard Stansbury. Philadelphia, 1852.

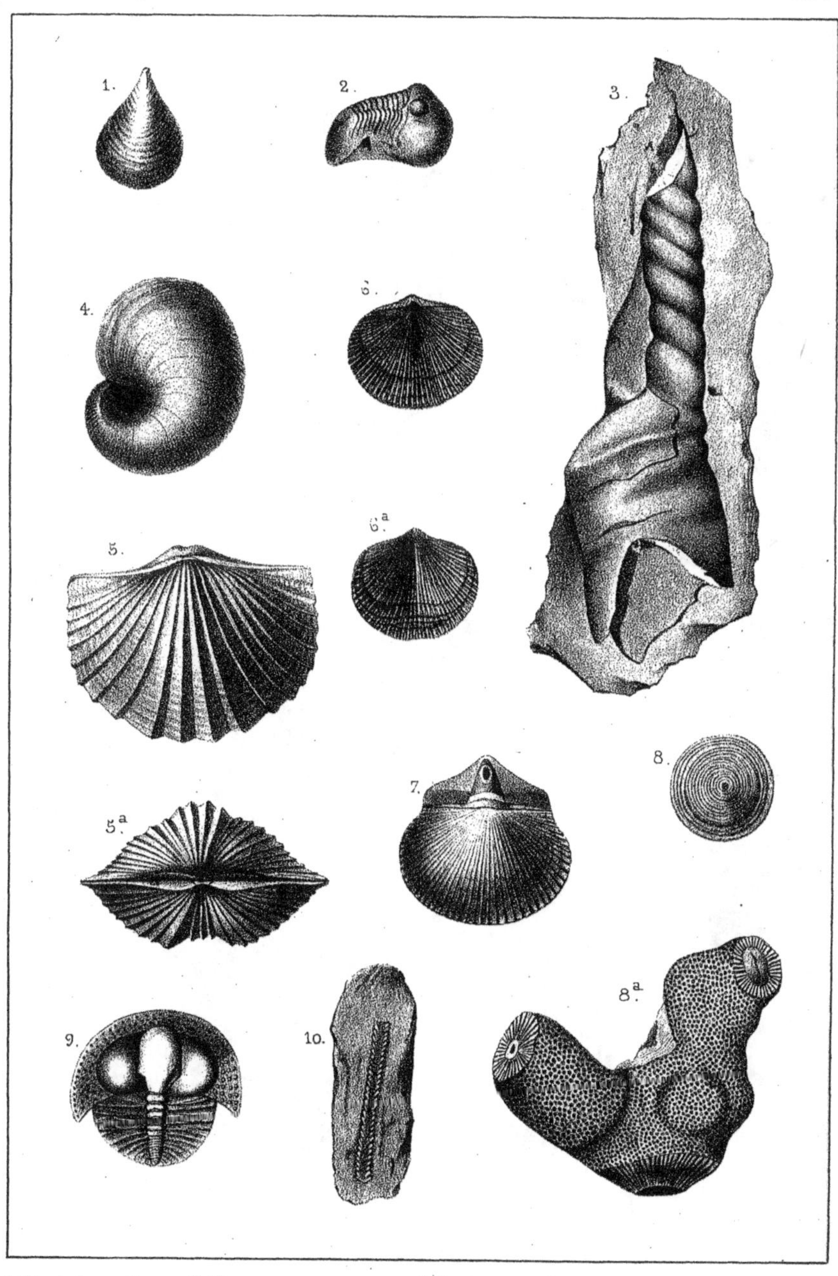

1. Lingula antiqua, Hall.
2. Illænus crassicauda, Wahl.
3. Orthoceratites communis, Wahl.
4. Bellerophon bilobatus, Murch.
5. Spirifer lynx, Eichw.
6. Orthis testudinaria, Dalm.
7. O. Verneuili, Eichw.
8. Chætetes Petropolitanus, Pand.
9. Trinucleus Caractaci, Murch.
10. Graptolites pristis, Hall.

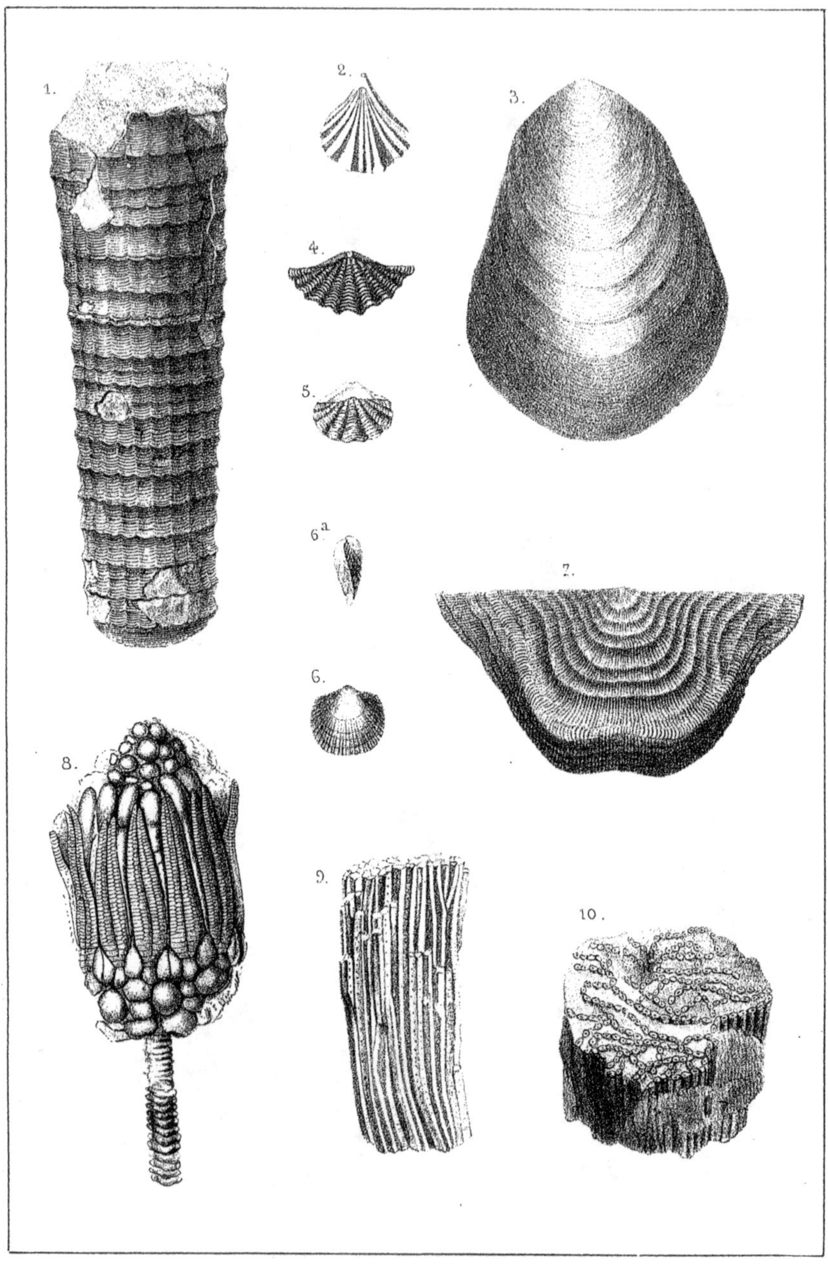

1. Orthoceratites annulatus, Sow.
2. Terebratula cuneata, Dalm.
3. Pentamerus oblongus, Sow.
4. Spirifer sulcatus, Dalm.
5. S. crispus, Dalm.
6. Orthis hybrida, Sow.
7. Leptæna depressa, Sow.
8. Hypanthocrinites decorus, Phil.
9. Favorites Gothlandica, Goldf.
10. Catenipora escaroides, Lamk.

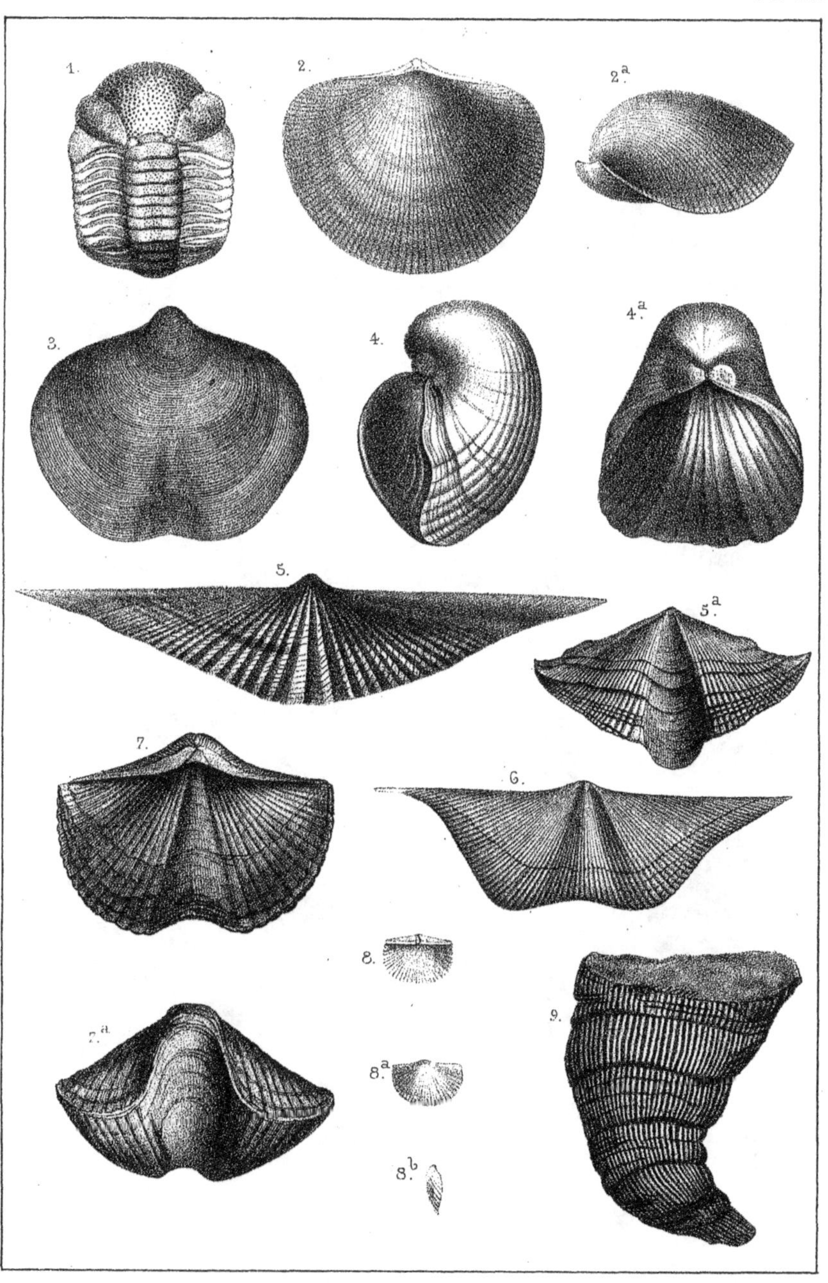

1. Calymene bufo, Green.
2. Terebratula reticularis, Linn.
3. T. concentrica, de Buch.
4. Pentamerus galeatus, Dalm.
5. Spirifer mucronatus, Conr.
6. S. Verneuili, Murch.
7. S. heteroclitus, Defr.
8. Chonetes nana, de Vern.
9. Zaphrentis gigantea, Raf.

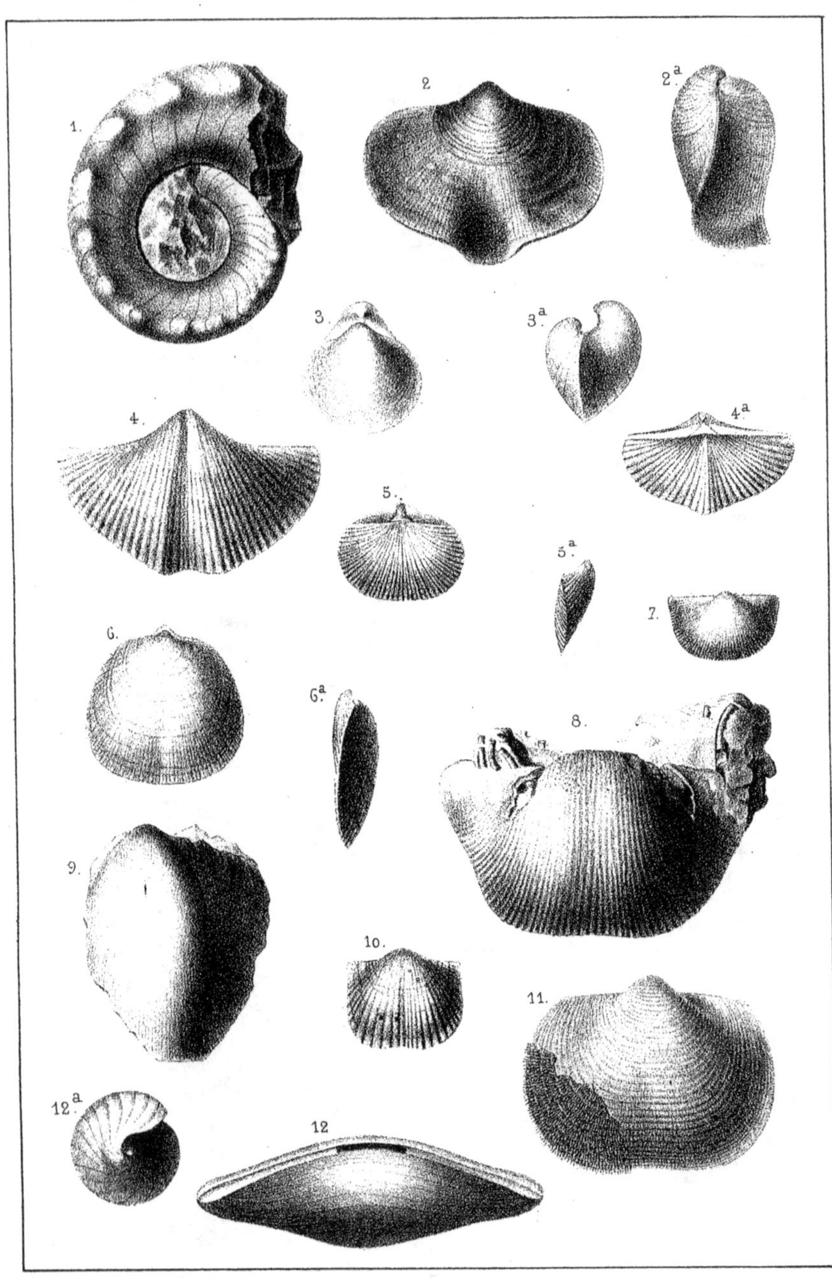

1. Nautilus tuberculatus, Sow.
2. Terebratula Roissyi, Lév.
3. Spirifer lineatus, Mart.
4. S. striatus, Mart.
5. Orthis crenistria, Phill.
6. O. Michelini, Lév.
7. Chonetes sarcinulata, Schlot.
8. Productus semireticulatus, Mart.
9. P. Cora, d'Orb.
10. P. Flemingi, Vern.
11. P. punctatus, Sow.
12. Fusulina cylindrica, Fisch.

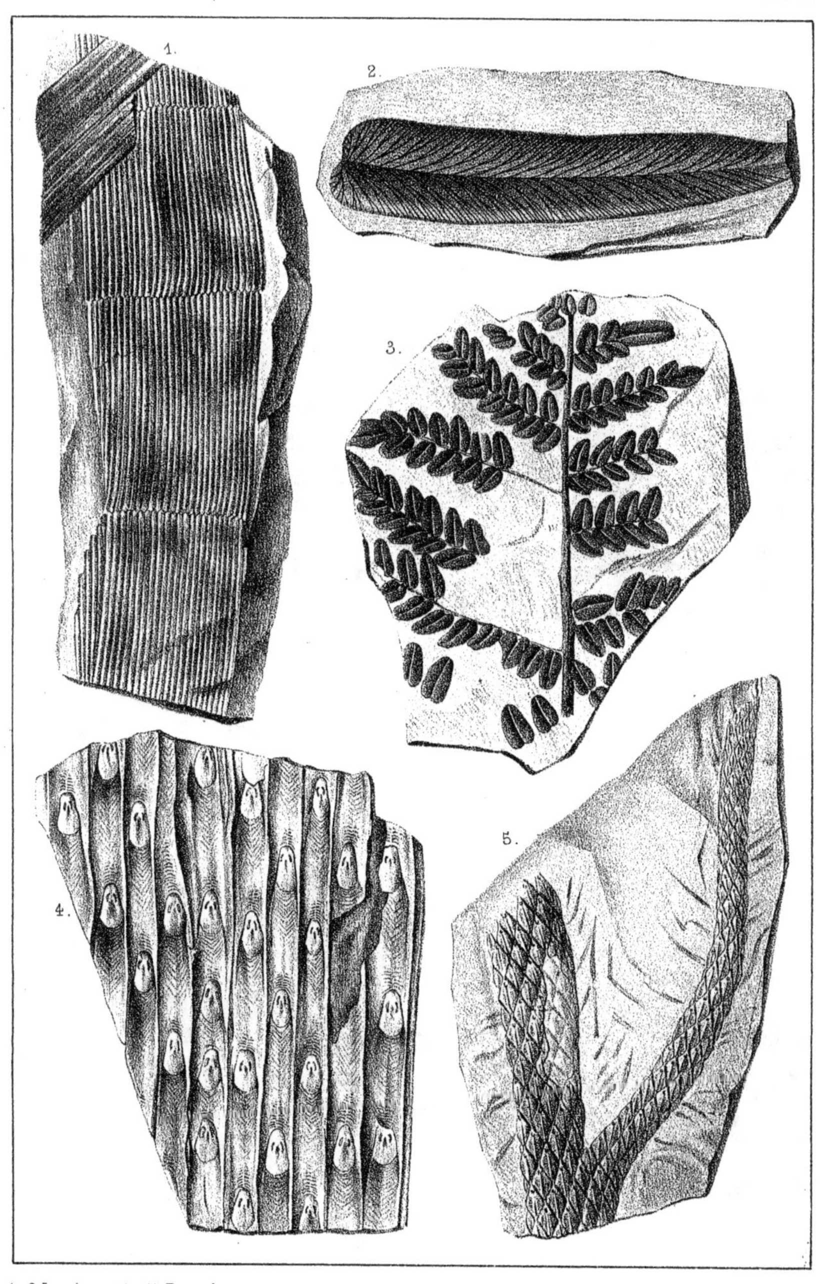

1. Calamites Cistii, Brong.
2. Nevropteris Angustifolia, Brong.
3. Nevropteris Loshii, Brong.
4. Sigillaria Sillimani, Brong.
5. Lepidodendron elegans, Brong.

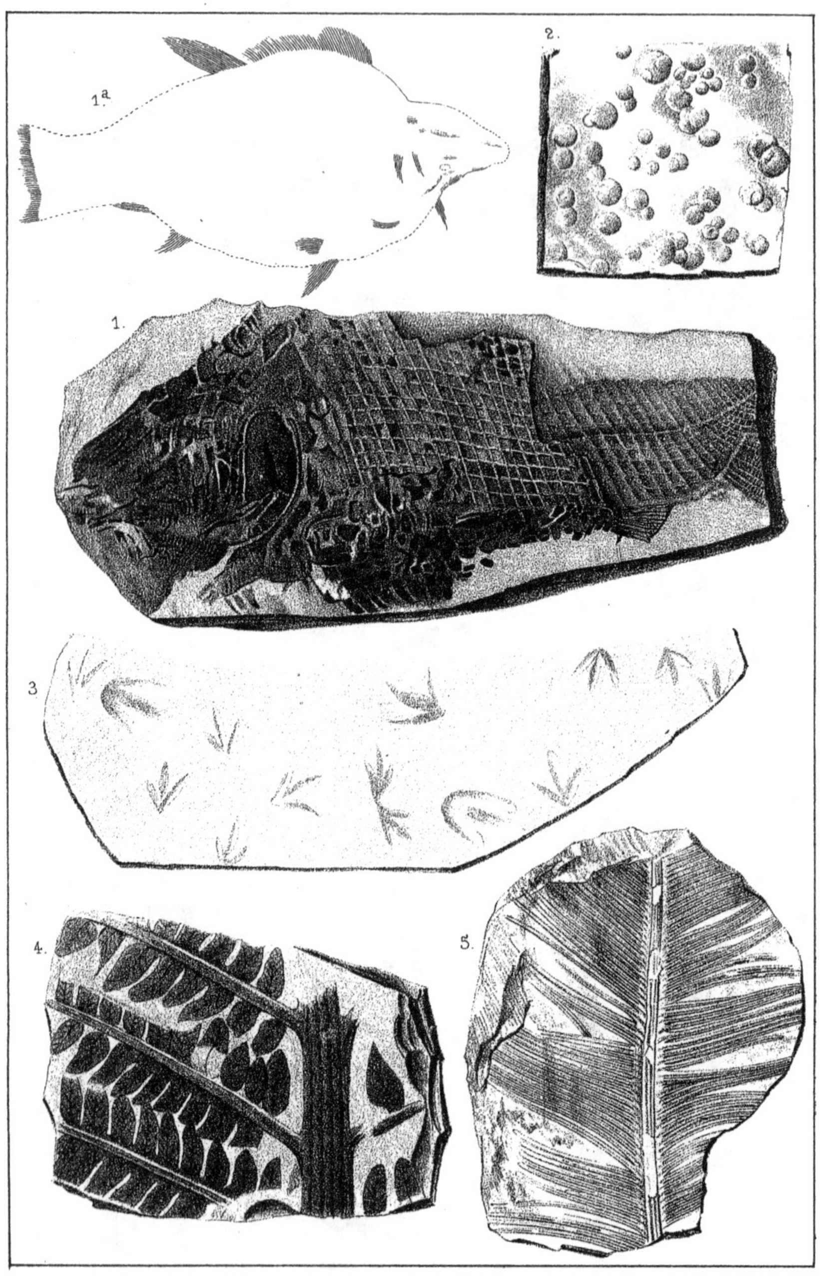

1. Eurynotus tenuiceps, Ag.
2. Impressions of rain drops, Redfield.
3. Ornithoidichnites tuberosus, Hitch.
4. Pecopteris Whitbiensis, Brong.
5. Zamites vittarioides, Brong.

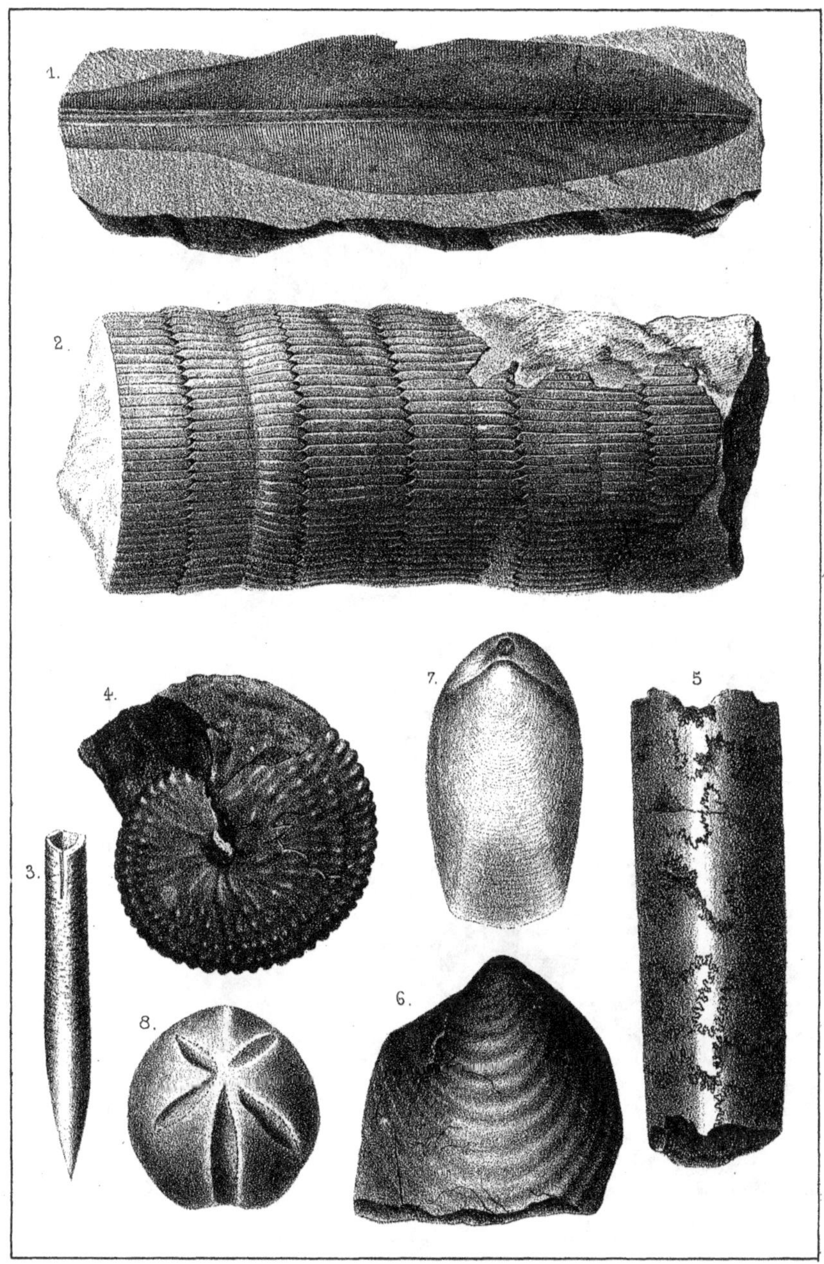

1. Teniopteris magnifolia, Rogers.
2. Equisetum columnare Brong.
3. Belemnitella mucronata, Blainv.
4. Ammonites Nebrascensis, Owen.
5. Baculites ovatus, Say.
6. Inoceramus Sagensis, Owen.
7. Terebratula Harlani, Mort.
8. Hemiaster parastatus, Mort.

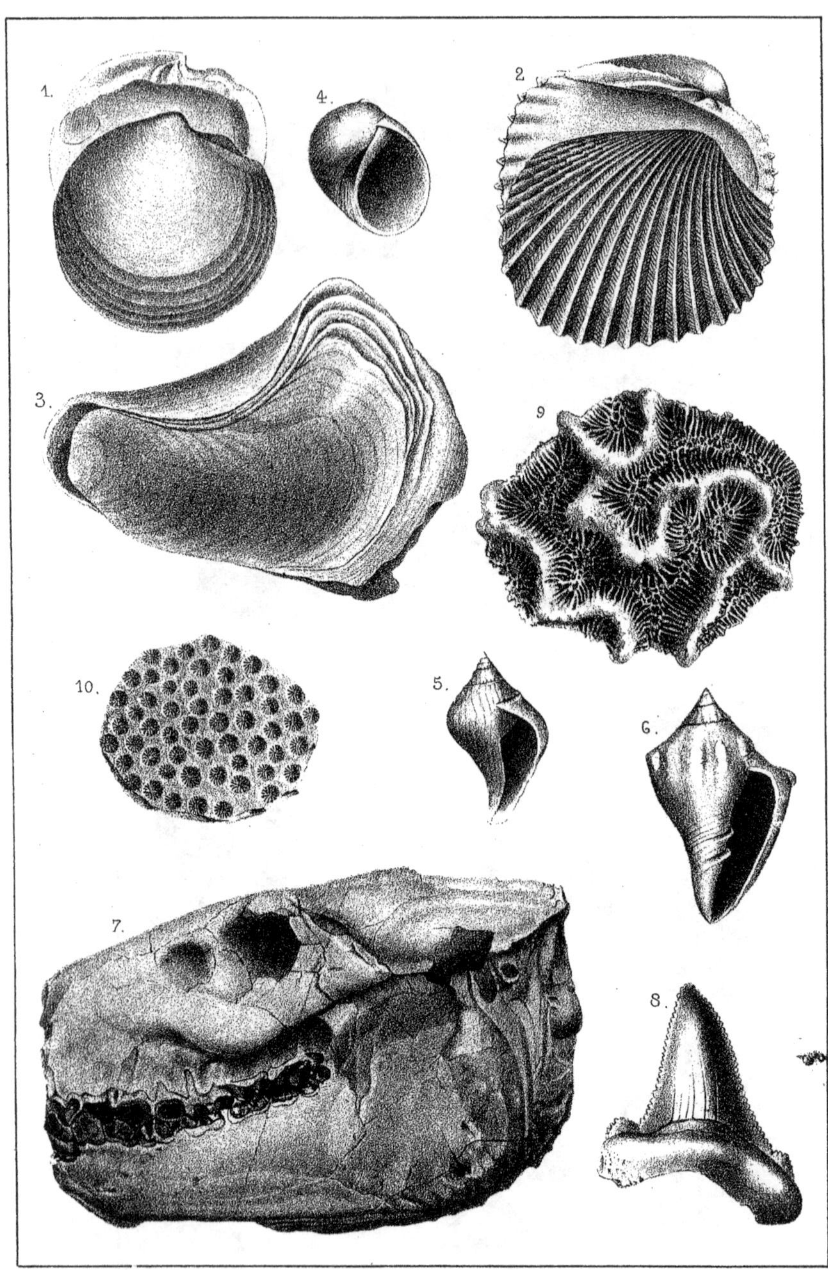

1. Lucina rotunda, Lea.
2. Venericardia Sillimani, Lea.
3. Ostrea semilunata, Lea.
4. Natica striata, Lea.
5. Fusus Fittoni, Lea.
6. Voluta Vanuxemi, Lea.
7. Oreodon Culbertsonii, Leidy
8. Carcharodon angustidens, Ag.
9. Meandrina labyrintica, Lamk.
10. Astrea mamillata, Lamk.

VALUABLE SCIENTIFIC WORKS.

THE FOOT-PRINTS OF THE CREATOR; or, the Asterolepsis of Stromness, with numerous illustrations. By HUGH MILLER, author of "The Old Red Sandstone," &c. From the third London Edition. With a Memoir of the author, by LOUIS AGASSIZ. 12mo,..cloth,....1,00

DR. BUCKLAND, at a meeting of the British Association, said he had never been so much astonished in his life, by the powers of any man, as he had been by the geological descriptions of Mr. Miller. That wonderful man described these objects with a facility which made him ashamed of the comparative meagreness and poverty of his own descriptions in the "Bridgewater Treatise,' which had cost him hours and days of labor. *He would give his left hand to possess such powers of description as this man;* and if it pleased Providence to spare his useful life, he, if any one, would certainly render science attractive and popular, and do equal service to theology and geology.

"Mr. Miller's style is remarkably pleasing; his mode of popularizing geological knowledge unsurpassed, perhaps unequalled; and the deep reverence for Divine Revelation pervading all, adds interest and value to the volume."—*New York Com. Advertiser.*

"The publishers have again covered themselves with honor, by giving to the American public, with the Author's permission, an elegant reprint of a foreign work of science. We earnestly bespeak for this work a wide and free circulation, among all who love science much and religion more."—*Puritan Recorder.*

THE OLD RED SANDSTONE; or, New Walks in an Old Field. By HUGH MILLER Illustrated with Plates and Geological Sections. 12mo,..................cloth,....1,00

"Mr. Miller's exceedingly interesting book on this formation is just the sort of work to render any subject popular. It is written in a remarkably pleasing style, and contains a wonderful amount of information."— *Westminster Review.*

"It is withal, one of the most beautiful specimens of English composition to be found, conveying information on a most difficult and profound science, in a style at once novel, pleasing and elegant. It contains the results of twenty years close observation and experiment, resulting in an accumulation of facts, which not only dissipate some dark and knotty old theories with regard to ancient formations, but establish the great truths of geology in more perfect and harmonious consistency with the great truths of revelation." — *Albany Spectator.*

PRINCIPLES OF ZOOLOGY: Touching the Structure, Development, Distribution, and Natural Arrangement of the RACES OF ANIMALS living and extinct, with numerous illustrations. For the use of Schools and Colleges. Part I., COMPARATIVE PHYSIOLOGY. By LOUIS AGASSIZ and AUGUSTUS A. GOULD. Revised edition. 12mo,...cloth,....1,00

"This work places us in possession of information half a century in advance of all our elementary works on this subject. * * No work of the same dimensions has ever appeared in the English language, containing so much new and valuable information on the subject of which it treats."—*Prof. James Hall, in the Albany Journal.*

"A work emanating from so high a source hardly requires commendation to give it currency. The volume is prepared for the *student* in zoological science; it is simple and elementary in its style, full in its illustrations, comprehensive in its range, yet well condensed, and brought into the narrow compass requisite for the purpose intended."—*Silliman's Journal.*

"The work may safely be recommended as the best book of the kind in our language."—*Christian Examiner.*

"It is not a mere book, but a work—a real work in the form of a book. Zoology is an interesting science, and here is treated with a masterly hand. The history, anatomical structure, the nature and habits of numberless animals, are described in clear and plain language and illustrated with innumerable engravings. It is a work adapted to colleges and schools, and no young man should be without it."—*Scientific American.*

PRINCIPLES OF ZOOLOGY, PART II. Systematic Zoology, in which the Principles of Classification are applied, and the principal groups of animals are briefly characterized. With numerous illustrations. 12mo,..................[in preparation.

VALUABLE SCHOOL BOOKS.

ELEMENTS OF MORAL SCIENCE, by FRANCIS WAYLAND, D. D., President of Brown University, and Professor of Moral Philosophy. Forty-seventh thousand.—12mo,..cloth,....1,25

MORAL SCIENCE ABRIDGED, and adapted to the use of Schools and Academies, by the Author. Thirtieth thousand,..................................half mor.... ,50

The same, CHEAP SCHOOL EDITION,....................................boards,.... ,25

This work is used in the Boston Schools, and is exceedingly popular as a text book wherever it has been adopted.

ELEMENTS OF POLITICAL ECONOMY, by FRANCIS WAYLAND, D. D. Twenty-first thousand. 12mo,..cloth,... 1,25

POLITICAL ECONOMY ABRIDGED, and adapted to the use of Schools and Academies, by the Author. Seventh thousand,......................half mor.... ,50

The above works by Dr. Wayland, are used as Text Books in most of the Colleges and higher Schools throughout the Union, and are highly approved.

PALEY'S NATURAL THEOLOGY. Illustrated by forty Plates, with selections from the Notes of Dr. Paxton, and additional Notes, original and selected, with a Vocabulary of Scientific Terms. Edited by JOHN WARE, M. D. 12mo.....half mor.....1,25

ROMAN ANTIQUITIES AND ANCIENT MYTHOLOGY; by C. K. DILLAWAY. Illustrated by elegant Engravings. Eighth edition, improved. 12mo..half mor.... ,67

THE YOUNG LADIES' CLASS BOOK; a Selection of Lessons for Reading, in Prose and Verse. By EBENEZER BAILEY, A. M. Fifty-second edition,..half mor.... ,84

BLAKE'S NATURAL PHILOSOPHY; being Conversations on Philosophy, with Explanatory Notes, Questions for Examination, and a Dictionary of Philosophical Terms, with twenty-eight steel Engravings. By J. L. BLAKE, D. D.,. ....sheep.... ,67

BLAKE'S FIRST BOOK IN ASTRONOMY; designed for the use of Common Schools. Illustrated with steel-plate Engravings. By JOHN L. BLAKE, D. D,............ half bound..... ,50

FIRST LESSONS IN INTELLECTUAL PHILOSOPHY; or a Familiar Explanation of the Nature and Operations of the Human Mind. By SILAS BLAISDALE,......... sheep,.... ,84

THE CICERONIAN; or, the Prussian Method of Teaching the Elements of the Latin Language. Adapted to the use of American Schools. By Professor B. SEARS, Secretary of Massachusetts Board of Education. 18mo,..............half mor.... ,50

MEMORIA TECHNICA; or, the Art of Abbreviating those Studies which give the greatest labor to the Memory; including Numbers, Historical Dates, Geography, Astronomy, Gravities, &c. By L. D. JOHNSON Second edition, revised and improved,........ half bound,...., .50

PROGRESSIVE PENMANSHIP, Plain and Ornamental, for the use of Schools. By N. D. GOULD, author of "Beauties of Writing," "Writing Master's Assistant," etc...... in five parts, each.... ,12½

LETTER SHEET SIZE of the above in four books,........... ...stiff covers, each.... ,20

The copies are arranged in progressive series, and are likewise so diversified by the introduction of variations in style, so as to command the constant attention and exercise the ingenuity of the learner, thus removing some of the most serious obstacles to the success of the teacher. They are divided into FIVE SERIES, intended for the like number of books, and are so arranged and folded that a copy always comes over the top of the page on which it is to be written.

There are ninety-six copies, presenting a regular inductive system of Penmanship for ordinary business purposes, followed by examples of every variety of Ornamental Writing.

☞ This work is introduced into many of the Boston Public and Private Schools, and gives universal satisfaction.

WRITING COPIES, Plain and Ornamental, from the "Progressive Penmanship," bound in one book.. ,16⅔

www.ingramcontent.com/pod-product-compliance
Lightning Source LLC
LaVergne TN
LVHW011120110826
845150LV00008B/2196

* 9 7 8 1 4 2 5 5 0 7 6 3 3 *